Topics in Applied Physics Volume 38

Topics in Applied Physics Founded by Helmut K. V. Lotsch

Charge-Coupled Devices

Edited by D.F. Barbe

With Contributions by
W.D. Baker   D.F. Barbe   H.K. Burke
K.L. Davis   J.M. Killiany   G.J. Michon
D.K. Schroder

With 133 Figures

Springer-Verlag Berlin Heidelberg GmbH 1980

David F. Barbe, Ph.D.

Office of ASN (R, E&S), Room 5E787, The Pentagon, Washington, DC 20350, USA

ISBN 978-3-662-31270-4   ISBN 978-3-540-38985-9 (eBook)
DOI 10.1007/978-3-540-38985-9

Library of Congress Cataloging in Publication Data. Main entry under title: Charge-coupled devices. (Topics in applied physics; v. 38). Includes bibliographical references and index. 1. Charge-coupled devices. I. Barbe, David F., 1939–. II. Baker, Wilford Dean, 1942–. TK 7871.99.C45C44   621.3815'2   79-27421

2153/3130-543210

Preface

From the time the charge-coupled device (CCD) was invented until about 1974, the research effort on CCDs was broad, and the progress was rapid. In 1975, Sequin and Tompsett authored a book entitled *Charge Transfer Devices*. This book was published at the end of the CCD research period, and it proved to be exceedingly useful since it served as a summary of most of the basic theory of CCDs. In 1977, a chapter by Barbe and Campana entitled "Imaging Arrays Using the Charge-Coupled Concept" in the book *Advances in Image Pickup and Display*, edited by Kazan, reviewed the uses of CCDs in visible imaging applications.

Since these works were published, the field of CCDs has progressed farthest in the areas of charge-injection devices (CIDs) for visible imaging applications, CIDs and CCDs for infrared applications, analog and digital signal processing devices and radiation hardening of CCDs. Thus, the need for this book. This book, then, taken with the works referenced above should provide a rather complete and up-to-date treatment of the field.

Other works referenced often in this book are papers in the proceedings of the topical conferences in CCDs. The first such conference was a national conference sponsored by the U.S. Navy and held in San Diego, California, in October 1973. Thereafter from 1974 to 1976, the conference was international in character and was held at the University of Edinburgh in even-numbered years and in San Diego in odd-numbered years. By 1976, most of the physics in the CCD area was known and what remained was engineering; thus, startling discoveries and innovations became fewer in number. As a result no conference was held in 1977 but enough progress had been made by 1978 so that a conference in San Diego was held.

The authors of the chapters in this book, workers deeply involved in the research, development, and application of CCDs, have made good use of the wealth of information in the proceedings of these conferences. Thus, this book should prove to be timely since it summarizes the development phase of CCDs as the book by Tompsett and Sequin summarized the research phase.

The book is organized in the following way: the first part, comprised of three chapters, discusses visible and infrared imaging with CCDs and CIDs; the second part, comprised of one chapter, discusses signal processing with CCDs; and the third part discusses radiation effects on CCDs and CIDs.

Washington, DC, November, 1979   *D. F. Barbe*

Contents

Contributors

Baker, Wilford D.
   Northrop Corporation, Northrop Research and Technology Center,
   One Research Park, Palos Verdes Peninsula, CA 90274, USA

Barbe, David F.
   Office of ASN (R, E&S), Room 5E787, The Pentagon,
   Washington, DC 20350, USA

Burke, Hubert K.
   General Electric Company, Research and Development Center
   Schenectady, NY 12301, USA

Davis, Kenneth L.
   Naval Research Laboratory, Washington, DC 20375, USA

Killiany, Joseph M.
   Code 5213 Naval Research Laboratory, Washington, DC 20375, USA

Michon, Gerald J.
   General Electric Company, Corporate Research and Development
   Schenectady, NY 12301, USA

Schroder, Dieter K.
   Westinghouse Research and Development Center, 1310 Beulah Road
   Pittsburgh, PA 15235, USA

1. Introduction

D. F. Barbe

1.1 Application Areas

The charge-coupled device (CCD) concept was invented in 1969 by *Boyle* and *Smith* [1.1]. They were searching for an equivalent to the magnetic bubble memory which could be fabricated in silicon and thereby make use of the well-developed silicon technology. While CCDs have made quite an impact in the memory field, they perform two other generic functions, i.e., imaging and signal processing.

In the field of imaging, CCDs have had a significant impact. The CCD provides a way to read out an electrical replica of a visible scene without introducing significant fixed pattern or temporal noise. This has provided a significant improvement in the state of the art of visible staring imaging [1.2].

Very early in the development of CCD technology it was recognized that CCDs could play a significant role in advanced infrared systems [1.3]. There are a number of functions which CCDs can perform in infrared focal planes, e.g., detection, readout, multiplexing, and time delay and integration (TDI). In scanned ir systems, TDI is one of the most important functions performed by CCDs. In such a system, the scene is mechanically scanned across an array of detector elements. By using CCD columns to shift the detector output signals (in the form of charge packets) along the focal plane with the same speed as the mechanical scan moves the scene across the array, the signal-to-noise ratio can be improved by the square root of the number of detector elements in the TDI column. At the systems level, such an increase provides for improved performance, e.g., increased range. Virtually every advanced focal plane design uses CCDs for one or more of the functions mentioned above [1.4].

Since the CCD can shift charge packets representing samples of analog signals, it can perform several analog and digital signal processing functions. Among these functions are: delay, multiplexing, demultiplexing, transversal filtering, recursive filtering, time delay and integration, analog memory, digital memory, and digital logic. With this variety of functions, CCDs are being used widely in special applications requiring a high degree of functionality with small size and low power [1.5].

1.2 Scope of this Volume

The purpose of this volume is to summarize the development phase of CCDs. It consists of three parts: the first part consists of three chapters treating visible and infrared imaging with CCDs and CIDs; the second part consists of one chapter treating signal processing, and the third part consists of one chapter treating radiation effects in CCDs and CIDs.

The chapter of Michon and Burke on CID image sensing provides an up-to-date summary of the design and operation of CIDs in visible imaging applications. This chapter represents the first summary of CID imaging to be included in a book.

The chapters of Baker and Schroder on infrared imaging are the first comprehensive articles on infrared applications of CCDs and CIDs to be included in a book. The chapter by Baker discusses the uses of CCD and CID structures which utilize narrow band gap (intrinsic) semiconductors. The chapter by Schroder discusses the uses of silicon, doped with impurities whose positions in the energy gap provide an infrared response, in CCD structures.

The chapter by Barbe, Baker, and Davis on CCD signal processing provides an up-to-date summary of the development of important CCD structures for signal processing applications.

The final chapter on radiation effects in CCDs and CIDs by Killiany provides the first comprehensive summary of this field to be included in a book. This work is especially important since many of the applications of CCDs and CIDs are in outer space where natural radiation causes a degradation in performance of metal oxide semiconductor devices. Another application where radiation effects are important is the use of CCD or CID imagers as monitoring devices in nuclear plants. Finally, the effects of nuclear radiation on CCDs and CIDs are often similar to the effects of other types of stress such as temperature-bias stress. Therefore, failure modes discovered from nuclear radiation stress often provide valuable information on the reliability of the devices.

1.3 Outlook

Early in the 1970s it was thought that CCDs would displace imaging tubes as visible image pickup devices. To date this has not happened except in certain specialized applications. There are two reasons for this – defects and cost. While many companies have made large-area TV-compatible CCD arrays, most have blemishes, defects which detract from or prevent their commercial use. The second factor (not independent from the first) is the cost of the arrays. While CCDs are integrated circuits and as such are mass producible using relatively standard silicon wafer manufacturing technology, the production volume must be large in order to accrue the benefits of mass production. But the cost must be low in order to attract a large-volume application. Although the present situation is relatively immobile, there is considerable potential for the future. Assuming that procedures will be found which will allow the dark current defect density to be held to tolerable levels, CCD and CID arrays should be attractive for use in high-volume applications such as the home video market. This application would use a CCD camera as the pickup device in a home movie system. The video would be stored via a video tape recorder. The TV set would be the display unit. This system would replace the film movie camera and movie projector presently used by most families. Obviously the prices of the video recorder and the CCD camera must be significantly lower than today's prices for a large market to develop for this system. Nevertheless, a huge market will exist when the price is right.

The applications of ir imaging are mostly for sophisticated noncommercial systems. In such instances, performance is the key factor and the cost of the CCD or CID arrays is usually a small part of the cost of the overall system. CCDs and CIDs have already had a significant impact on this application area. Virtually every advanced ir system will use CCDs or CIDs in some way.

The applications of CCDs as sampled data signal processing elements are rather specialized and varied. These devices are not now commercially available and will probably not be in the future because each application requires a specific design. Therefore, sampled data devices will probably continue to be custom-designed components for specific applications where size, power, functionality, and cost are important factors.

References

1.1 W. S. Boyle, G. E. Smith: Bell Syst. Tech. J. **49**, 587–593 (1970)
1.2 D. F. Barbe, S. B. Campana: "Imaging Arrays Using the Charge-Coupled Concept", in *Advances in Image Pickup and Display*, Vol. 3, ed. by B. Kazan (Academic Press, New York 1977)
1.3 D. F. Barbe, W. A. Schmidt: "Infrared Charge-Coupled Devices – Problems, Progress and Prognosis", in 1972 IRIS Imaging Specialty Group Proc., pp. 1–25
1.4 D. F. Barbe: Electro-Opt. Syst. Des. **9**, 50–58 (1977)
1.5 D. F. Barbe, W. D. Baker, K. L. Davis: IEEE Trans. ED-**25**, 108–125 (1978)

2. CID Image Sensing

G. J. Michon and H. K. Burke

With 22 Figures

The charge-injection device (CID) is a surface channel device that employs intracell charge transfer and charge injection to achieve the solid-state image sensing function. Photon-generated charge signals are collected and stored in an array of MOS (metal-oxide semiconductor) charge storage capacitors. The level of signal charge is detected in situ so that excess charge transfer structure is avoided. Charge injection into the underlying semiconductor is used to clear the sensing region of accumulated signal charge and, in some cases, to provide a readout means.

2.1 Basic Operation

The charge-injection approach to solid-state image sensing employs MOS capacitor structures to collect and store photon-generated charge signals. Charge is injected from the MOS storage (inversion) region into the substrate to clear the storage region and, in some cases, to provide a signal readout means. Charge storage sites can be designed for linear addressing (line imager) or for two-dimensional coincident voltage addressing (area imager).

2.1.1 Linear Structure

The simplest charge-injection device is the MOS capacitor. If voltage is applied to an MOS capacitor, the underlying silicon is depleted of majority carriers and photon-generated minority carriers can be collected and stored in a surface inversion region as illustrated in Fig. 2.1. If voltage is subsequently removed from the capacitor, the stored charge will be injected into the substrate where it can either recombine or be removed by a charge-collection structure.

The magnitude of the injected charge can be determined by measuring the displacement current that flows in the external circuit. If the displacement current that flows in the substrate (and gate) circuit is integrated over the injection time interval, the waveform shown in Fig. 2.1d is obtained. The difference between the levels before and after injection is proportional to the net injected charge.

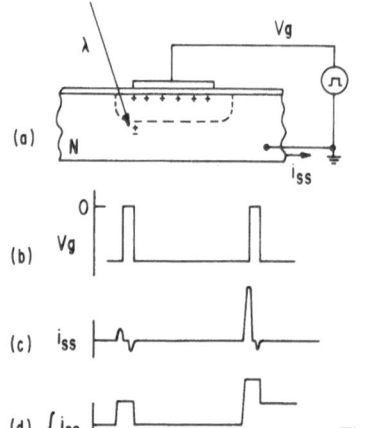

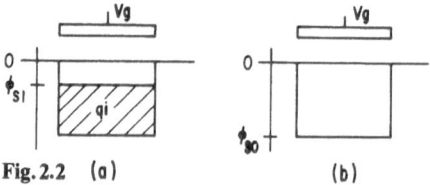

Fig. 2.2 (a)   (b)

Fig. 2.1a–d. MOS capacitor operated in the charge-injection mode (a) device cross section, (b) drive voltage, (c) displacement current, and (d) integrated displacement current, without and with injected charge, respectively

Fig. 2.2a, b. Semiconductor surface potential and stored charge (a) before and (b) after injection

Fig. 2.1

2.1.2 Depletion Capacitance Loading

The charge signal in a CID is measured at the MOS storage capacitor where it was collected. The semiconductor depletion capacitance consequently is a part of the readout circuit and has a loading effect on the output signal. This effect can be seen for the injection readout of Fig. 2.1 by considering the net change in circuit charge upon injection. The diagram of Fig. 2.2 illustrates the silicon surface potential and inversion layer charge before and after injection. Neglecting interface states and fixed oxide charge, the charge stored on the oxide capacitance q_{ox} must equal the sum of the charge in the semiconductor depletion region q_d and in the inversion layer q_i. The space charge in the depletion region is equal to the product of the depleted volume and the doping density. For a relatively large storage capacitor in which lateral depletion can be neglected, a one-dimensional analysis can be used to calculate the depletion region charge. The depletion width W is given by *Grove* [Ref. 2.1, p. 159] as

$$W = (2\varepsilon_s \phi_s / q N_B)^{1/2} \tag{2.1}$$

(ε_s: semiconductor dielectric constant, ϕ_s: semiconductor surface potential, q: electronic charge, and N_B: minority carrier concentration).
Depletion region charge

$$q_d = W N_B q = (2\varepsilon_s \phi_s q N_B)^{1/2}. \tag{2.2}$$

Oxide charge

$$q_{ox} = (V_g - V_{fb} - \phi_s) C_o \tag{2.3}$$

(V_g: applied voltage, V_{fb}: flat band voltage, and C_o: oxide capacitance).

$$q_{ox} = q_d + q_i. \tag{2.4}$$

The net charge that flows in the external circuit upon injection is equal to the change in surface potential multiplied by the oxide capacitance. For lightly doped silicon ($5 \times 10^{14}\,cm^{-3}$) the charge readout is approximately 94 % of the collected charge.

More typically, lateral depletion is not negligible. Figure 2.3 illustrates the volume of the space charge region under an electrode of dimensions X by Y. The loading imposed by the depletion capacitance is a function of electrode size. The charge readout from a storage capacitor as small as $130\,\mu m^2$ is greater than 80 % of the charge collected with this structure.

2.1.3 Interface States

When an MOS capacitor is biased above its threshold voltage, imperfections at the semiconductor-insulator interface act as traps for inversion layer charge. Trapped charge is immobile and cannot be transferred as required for a charge-transfer device. If the surface region is subsequently depleted of mobile charge, the trapped charge is released after a time interval that is related to the activation energy of the specific traps present [2.2]. This phenomenon is responsible for one component of charge-transfer loss in surface channel devices. If the storage capacitor is driven into accumulation, the surface states apparently act as recombination centers. Any charge that is trapped at the time that the region is driven into accumulation recombines and is lost. This action has been termed charge pumping by *Brugler* and *Jespers* [2.3].

The single MOS capacitor operated in the CID sensor mode does not require charge transfer for proper operation and can function with very high interface state densities. The CID sensing site that is used for two-dimensional area image sensing does require at least one charge transfer for proper operation and is somewhat sensitive to interface states.

2.1.4 X–Y Addressable Sensing Site

If two MOS capacitors are used at each sensing site and are coupled together so that stored charge can be transferred from one capacitor to the other, then a two-axis selection method can be used for scanning. The basic approach is to design each capacitor such that it can store the signal charge when voltage is removed from the other capacitor electrode. Injection will then occur only when both electrode voltages are switched off.

Various methods can be used to couple surface charge between adjacent electrodes. Among these are fringing fields, which require a very narrow interelectrode gap, overlapping but insulated electrodes, or the use of an

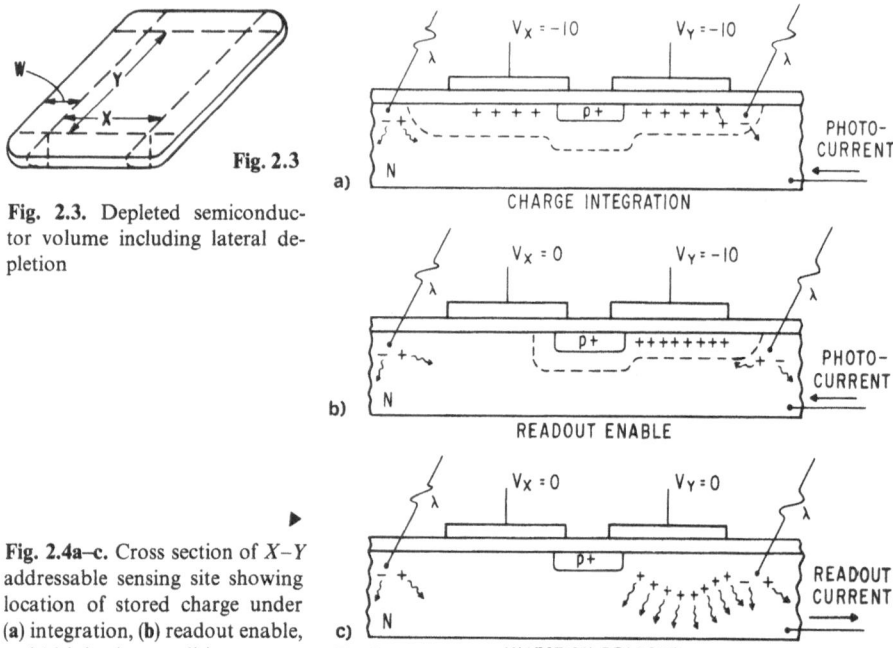

Fig. 2.3. Depleted semiconductor volume including lateral depletion

Fig. 2.4a–c. Cross section of $X–Y$ addressable sensing site showing location of stored charge under (a) integration, (b) readout enable, and (c) injection conditions

interconnecting conductive region. The last method is compatible with standard MOS processes and metal patterning capability and has been used for most of the CID imagers fabricated to date.

A very simple topological structure results with this image sensing technique. The MOS capacitors at each sensing site are coupled with a diffusion. There is no contact made to this diffusion. Array interconnections are readily made using the two-conductor level capability of self-registered MOS processes such as silicon gate.

Figure 2.4a shows the cross section of a sensing site with voltages applied to both electrodes. If voltage is removed from either of the electrodes (Fig. 2.4b), charge stored under that electrode will transfer to the other capacitor through the coupling region. Charge is injected only from the sensing site that has both electrode voltages switched off simultaneously (Fig. 2.4c). This arrangement permits coincident, two-dimensional $(X–Y)$ scanning, in any order. Sequential scanning can be implemented by including MOS shift registers along two edges of the array. Nonsequential ("random") scanning can be mechanized with digital decoders for row and column selection.

If array drive voltages are switched below the threshold voltage during half-select operation, charge pumping will cause a signal charge loss. A bias voltage, slightly greater than the threshold voltage, can be added to the drive levels to avoid this loss. A bias charge, similar to the CCD fat zero, will then accumulate at each site.

2.1.5 Charge Injection

Charge is removed from CID image sensors by injecting the minority carriers, which have been stored in surface inversion regions, into the underlying semiconductor. The first devices [2.4] were constructed on bulk silicon and relied on recombination as the primary method of charge removal. In this approach there is a tradeoff among site density, dark current, and cross talk. The diffusion length for minority carriers is proportional to the square root of carrier lifetime, while the thermal charge generation rate in depleted bulk silicon is inversely proportional to lifetime [Ref. 2.1, pp. 124, 174]. If the spacing between sensing sites is much less than the diffusion length, a portion of the charge injected at one site will be collected by adjacent sites with a resulting loss of resolution. In addition, the injection pulse width cannot be much shorter than carrier lifetime or else part of the injected charge will be recollected and result in image lag.

The solution to these problems has been to fabricate CID imagers on epitaxial wafers [2.5]. The epitaxial junction, which underlies the imaging array, acts as a buried collector for the injected charge. If the thickness of the epitaxial layer is less than, or comparable to, the spacing between sensing sites, the injected charge will be collected by the reverse-biased epi junction and injection cross talk is avoided. The rate at which charge injected at the surface is removed from the epitaxial layer can be analytically determined. The results of an approximate analysis are shown in Fig. 2.5. Diffusion of a quantity of charge, from the surface to the epitaxial depletion region boundry, has been calculated assuming low level injection and infinite carrier lifetime. Figure 2.5a shows the charge distribution in the epitaxial layer at various times after injection. Measured injection efficiency (percent of stored charge injected) as a function of injection pulse width is shown in Fig. 2.6 for bulk and epitaxial imagers.

The epitaxial collector also affects imager sensitivity. Part of the charge generated in the silicon between sensing sites can be collected by the *epi* junction instead of the storage capacitors. This is particularly true for long wavelength radiation which generates charge further from the imager surface.

2.2 Readout Methods

The signal charge stored in CID imaging arrays can be sensed by measuring either the charge that flows upon injection or the voltage change induced by charge transfer between the two storage capacitors that comprise the $X-Y$ addressable storage site. If charge is injected for readout then the readout process is destructive because the injection operation clears the sensing sites of signal charge. This readout method will be called injection readout. Sequential *injection* and *preinjection* are examples of this technique.

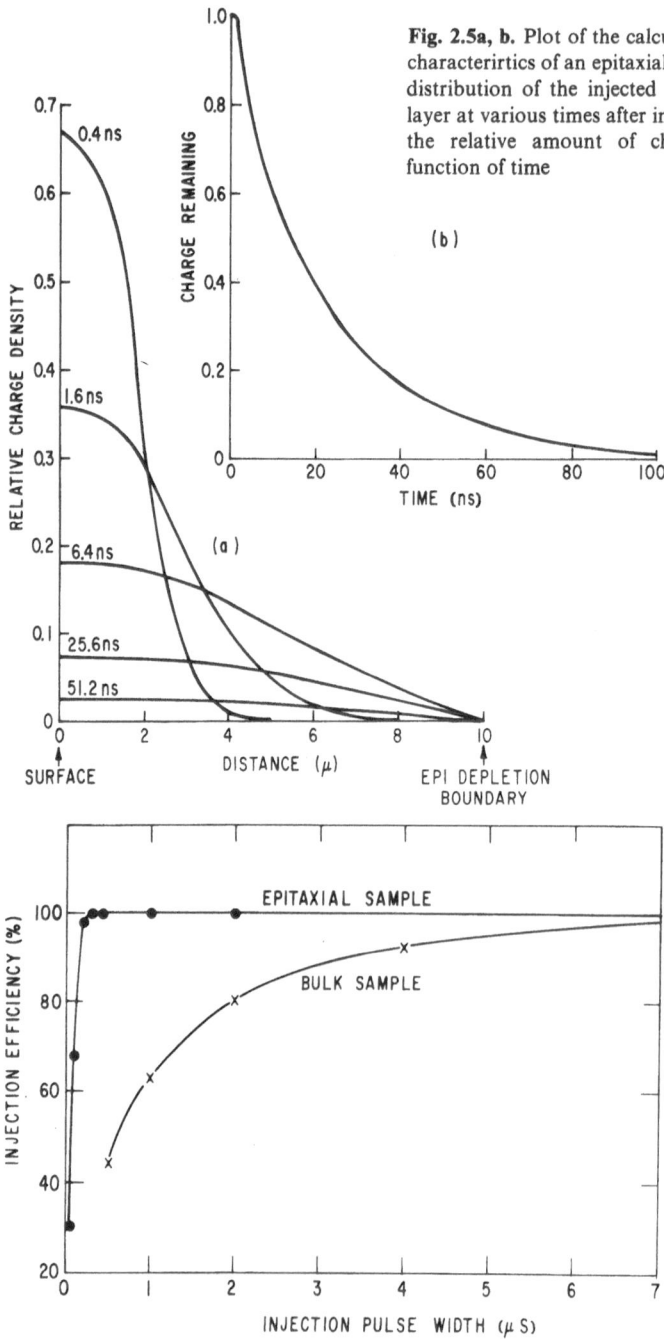

Fig. 2.5a, b. Plot of the calculated charge-collection characterirtics of an epitaxial junction. (**a**) Shows the distribution of the injected charge in the epitaxial layer at various times after injection, while (**b**) shows the relative amount of charge remaining as a function of time

INJECTION RESPONSE

Fig. 2.6. Measured injection efficiency (% of stored charge injected) of bulk and epitaxial imaging devices

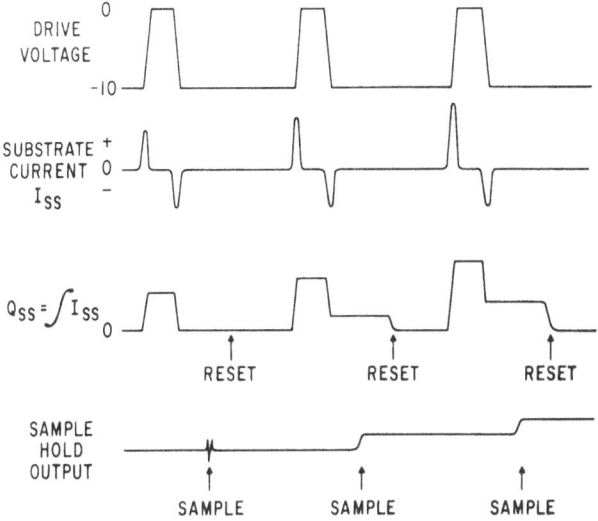

Fig. 2.7. Video signal waveforms for substrate readout illustrating column drive voltage, substrate current, net injected charge, and sampled video for different levels of injected charge

Readout can also be implemented by measuring the voltage induced by charge transfer between the two storage capacitors that are used at each sensing site in an array [2.6]. In an $X-Y$ addressable array the transfer can be performed on all sensing sites along a row in parallel. Each row can also be cleared of signal charge by performing the injection operation in parallel at all sites in the addressed row. This readout technique will be called transfer readout. It is nondestructive because the readout function has been separated from the injection operation. Image charge can be read out and retained or injected dependent upon the array drive voltage conditions. Parallel injection and row readout are examples of this second technique.

2.2.1 Sequential Injection

The initial charge-injection imagers utilized the substrate as a readout port common to all array sensing sites. With this approach, scanning can be implemented by first removing voltage from an array row (X line) and then pulsing each column (Y line) in sequence to readout the selected row. All rows can be read out in this manner, with or without interlace, dependent upon the order of row selection.

Video signal waveforms obtained using substrate readout are illustrated in Fig. 2.7. The raw video signal consists of the substrate charge injected from each sensing site in sequence. This signal appears as a displacement current in the presence of drive line interference that results from parasitic capacitive coupling of the drive voltage to the substrate. The drive voltage interference can be made to cancel itself through the use of an integrating readout technique. The first pulse time of Fig. 2.7 shows the substrate current signal that results

 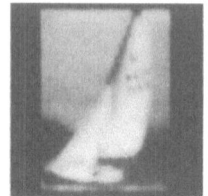

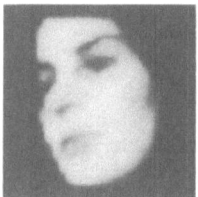 

Fig. 2.8. Images obtained from 32 × 32 self-scanned imagers employing substrate readout

from capacitive coupling of the Y line drive pulse in the absence of signal charge. The substrate current is simply $C(dV_Y/dt)$. If this current signal is integrated, the drive voltage waveform is recovered. The second pulse time shows the substrate current when signal charge is injected upon Y line drive voltage turnoff. The positive current pulse contains both the parasitic capacitance charge and the signal charge. The negative substrate current pulse that results from the reapplication of Y line drive voltage contains only the parasitic capacitance charge. The integral of this current waveform results in a net signal proportional to the injected signal charge. This net voltage is sampled to provide the video output voltage.

This signal recovery system results in self-cancellation of the parasitic drive line interference. For complete cancellation it is only necessary that the drive voltage return to its initial state – variations in the magnitude of the drive voltage or individual variations in line capacitance do not affect cancellation.

The signal voltage developed at the output of the image sensor is equal to the net injected charge divided by the total load capacitance. Capacitive loading imposed by the array lines can be minimized by allowing all but the selected row and column conductors to float during each line scan interval. If scanning circuitry is integrated into the imaging array then capacitance between the scanning shift registers and substrate will represent the bulk of the load capacitance. Images obtained from early 32 line by 32 element self-scanned imagers using substrate readout are shown in Fig. 2.8.

The displacement current that flows in the substrate upon charge injection also flows in the driven array line. The capacitive load of on-chip scanning circuitry can be avoided by sensing current in the driven line instead of the substrate.

An array designed for drive line readout which includes integral shift registers is diagrammed in Fig. 2.9a. A larger voltage is applied to the row

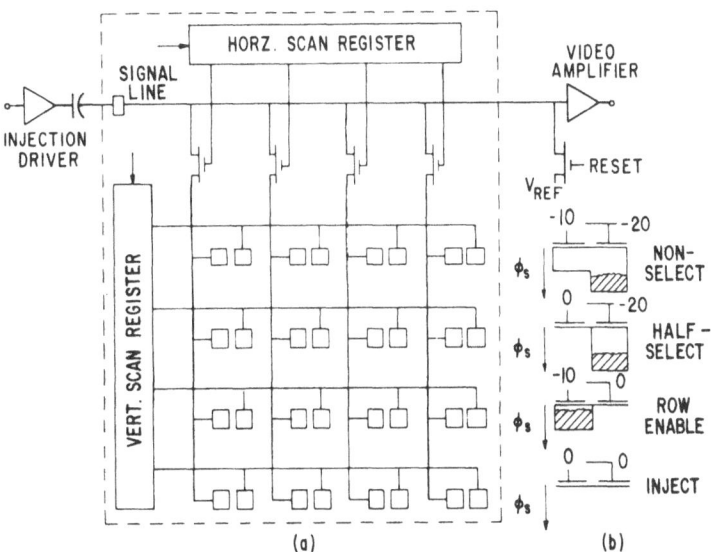

Fig. 2.9a, b. Diagram illustrating basic $X-Y$ accessing scheme for a CID imager. (a) Is a schematic diagram of a 4×4 array, while (b) shows the sensing site cross section showing silicon surface potentials and location of stored charge for various operating conditions

electrodes than to the column electrodes so that photon-generated charge collected at each site is stored under the row electrode thereby minimizing the capacitance of the column lines. The sensing site cross sections of Fig. 2.9b illustrate the silicon surface potentials and locations of stored charge under various applied voltage conditions.

A line is selected for readout by setting its voltage to zero by means of the vertical scan register. Signal charge at all sites of that line is transferred to the column capacitors, corresponding to the row enable condition shown in Fig. 2.9b. The charge is then injected by driving each column voltage to zero, in sequence, by means of the horizontal scan register and the signal line. The net injected charge is measured by integrating the displacement current in the signal line, over the injection interval. Charge in the unselected lines remains under the row-connected electrodes during the injection pulse time (column voltage pulse). This corresponds to the half-select condition of Fig. 2.9b.

One method for measuring the net injected charge is shown in Fig. 2.10. After column selection the signal line is reset and allowed to float. This reset operation introduces an uncertainty in the voltage on the signal line capacitance because of Johnson noise in the reset switch. This variation in the reset voltage is commonly called kTC noise and is further explained in Sect. 2.3.5 [2.15]. A sample of the signal after reset is taken to eliminate kTC noise by pulsing the RESTORE switch. The signal line is then driven by the capacitively coupled drive pulse to inject the signal charge at the selected imager site. After

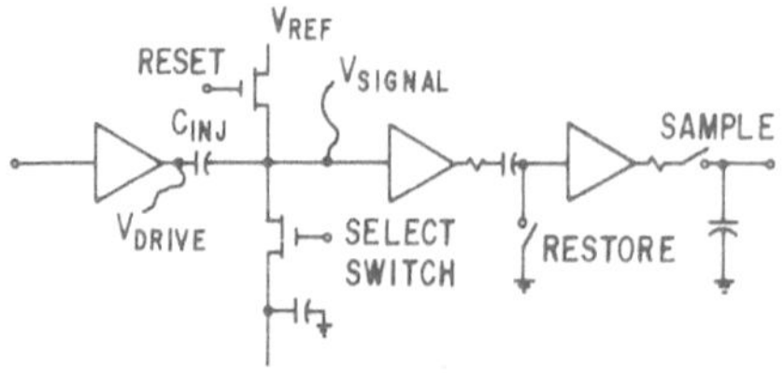

Fig. 2.10. Sequential injection imager readout

Fig. 2.11. Image obtained from a 100×100 self-scanned imager constructed on an epitaxial substrate and employing drive line sensing

▼

the drive voltage returns to its initial value, the net change in voltage on the signal line is proportional to the injected charge. This voltage is sampled to form the video signal.

The image obtained with a 100 line by 100 element array employing drive line sensing is shown in Fig. 2.11.

2.2.2 Preinjection

The preinjection readout technique is based upon the measurement of the change in charge that occurs at each addressed sensing site when a complete row of sites is cleared (injected) simultaneously. The schematic diagram of an array configured for preinjection readout is shown in Fig. 2.12. Equal row and column bias levels are normally used with this readout method. A low input impedance (transconductance) amplifier is used so that, during each scan of the array columns, the column potentials are reset to the reference voltage. Prior to each scan of the array columns, during the horizontal retrace interval, voltage is removed from the selected row to clear the row of sites to a bias charge level,

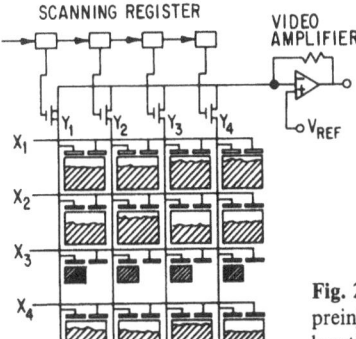

Fig. 2.12. Schematic diagram of a 4×4 CID array designed for preinjection readout. Silicon surface potentials and signal charge locations are indicated schematically

and then reset to its original value. The potential of all columns had been reset to the column reference potential during the previous scan interval. Since signal charge was present under the addressed row electrodes when the column potential was reset, the removal of the signal charge by the injection operation results in a voltage being induced on the floating column electrodes proportional to the injected signal charge. The induced video signal is then read out by the column scanner.

This preinjection readout method has a number of advantages and some limitations. Array fixed pattern noise is automatically rejected since the only net change in array charge levels prior to each video line scan is the injection of signal charge. The technique is compatible with high-speed sampling of the induced column signals as required for TV compatible operation. The main disadvantage with this readout technique is the switching noise coupled into the video signal by the column scanner. This component of fixed pattern noise repeats for each video line and can be rejected if one line of video storage is provided.

2.2.3 Parallel Injection

The parallel injection technique allows the functions of charge injection and charge detection to be separated. Signal charge levels can be sensed at high speed during a line scan, and during the line retrace time interval all of the charge in the selected line can be injected in parallel. If injection is deferred, nondestructive readout results. The parallel injection technique is well adapted to TV scan formats in that the signal is read out line by line. It is not adapted to random scan applications.

In an array of MOS coupled-capacitor pairs, as is used in the present charge-injection imagers, all of the signal charge will be stored under the row-connected electrodes if the row voltages are larger than the column voltages. This condition is illustrated in Fig. 2.13 for rows X_1, X_2, and X_4. This method of biasing effectively prevents the charge stored under the row-connected elec-

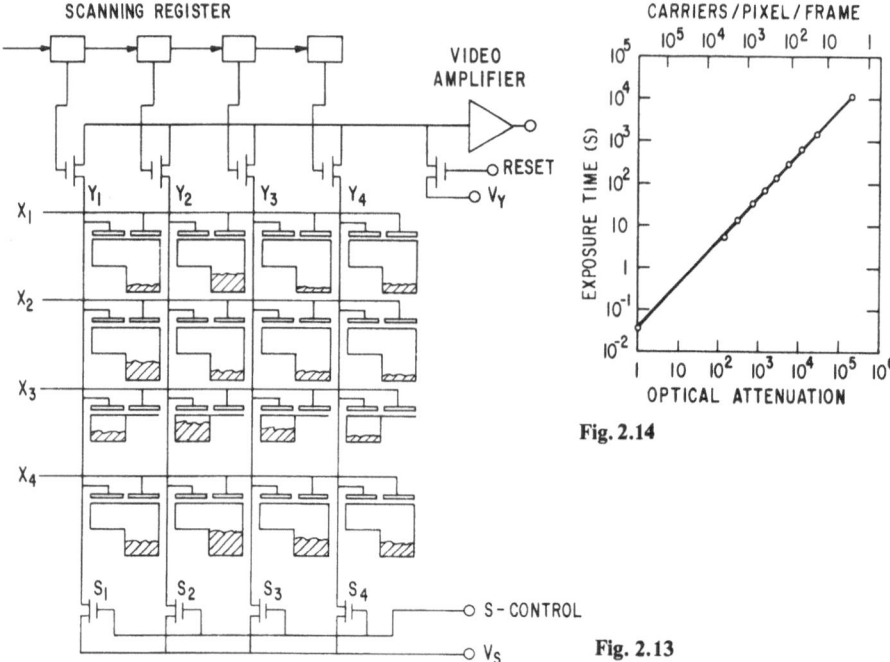

Fig. 2.14

Fig. 2.13

Fig. 2.13. Schematic diagram of a 4×4 CID array designed for parallel-injection readout. Silicon surface potentials and signal charge locations are indicated schematically

Fig. 2.14. Plot showing exposure time required to accumulate a given quantity of signal charge as a function of incident light level. Time is plotted as a function of optical attenuation. Measurements were made a 244×248 array, cooled to 200 K and operated at 30 frames per second, NDRO

trodes from affecting column voltages. The voltage on all array columns, Y_1 through Y_4 in Fig. 2.13, can be set to a reference value either by means of a previous column scan readout, or through the use of the column switches, S_1 through S_4.

If the voltage on a row electrode is then switched to zero, signal charge will transfer from the row-connected electrodes to the column-connected electrodes in the selected row of sensing sites. This is diagrammed in Fig. 2.13 for row X_3. The voltage on each of the column lines will then be reduced by an amount equal to the signal charge divided by the column capacitance.

The signal can be sensed by sequentially connecting each column line to a video amplifier by the use of a scanning register and MOS switches. The readout operation consists of resetting the video amplifier input to the reference voltage, and then stepping the scanning register to the next column line. After all columns of the array have been scanned, charge can be returned to the row-connected electrodes by reapplying voltage to the previously selected row. This action retains the signal charge for future processing, and constitutes a non-destructive readout (NDRO) operation.

Fig. 2.15. NDRO image readout

Alternately, at the end of readout of the selected row, while the row voltage is maintained at zero volts, the signal charge can be injected from the selected row to the substrate, all sites in parallel, by switching all column voltages to zero simultaneously. This action clears the sensing sites of charge and allows the start of a new signal integration time interval for that row.

Two experiments were performed to identify the precision to which the readout is nondestructive. First, the charge pattern of an image was generated and stored by momentarily opening a shutter, and then the image was read out continuously at 30 frames per second, until image degradation was noted. At a chip temperature of 200 K, images were read out for 3 h (324,000 NDRO operations) with no detectable charge loss. The charge lost during each NDRO operation was, on the average, much less than one carrier per pixel per frame.

The second experiment was performed to test whether charge could be generated and stored at very low light levels under continuous (30 frames per second) NDRO conditions. A series of time exposures was made at successivley lower light levels and the time required to reach a given level of signal voltage was measured. The results, Fig. 2.14, show that the exposure time is inversely proportional to light level with no measured charge loss for exposure times up to 3 h. The lowest light level used was equivalent to about two carriers per pixel per frame in the highlight regions of the image. Here again the readout loss was not detected and was much less than one carrier per pixel per frame.

Subsequent to the measurements described above, an image was stored on a 128 × 128 array for 65 h while being continuously read out nondestructively at 30 frames per second. The initial and final images are shown in Fig. 2.15.

2.2.4 Row Readout

A second charge-transfer readout method is diagrammed in Fig. 2.16. Readout is effected by driving the column connected electrodes to cause charge to transfer to the row connected electrodes. The condition diagrammed in Fig.

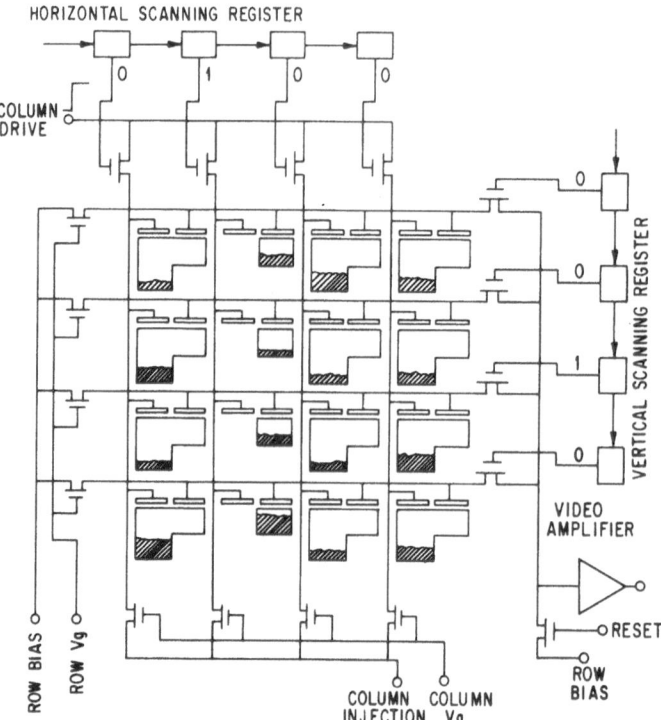

Fig. 2.16. Row readout array diagram illustrating sensing of the third row, second column

2.16 is with the third row connected to the video amplifier by the vertical scanning register. The horizontal scanning register is operated at the video element rate to sequentially connect the column drive voltage to the array columns. Each element of signal charge is transferred to the row electrode, externally sampled, and then transferred back to the column electrode.

At the end of each line scan, the selected row and all array columns can be driven to cause injection. For nondestructive readout of this array, the injection operation can be deferred.

The advantages of this readout method over previously described techniques is that the horizontal scanner is isolated from the video signal, kTC noise can be easily rejected, and nondestructive readout can be readily mechanized.

2.3 Performance Characteristics

Localized charge transfer and injection result in sensitivity, cross talk, and blooming characteristics that are different from other solid-state image sensing techniques.

2.3.1 Sensitivity

A cross-section sketch of the CID image sensing structure is shown in Fig. 2.17. Since signal charge is detected at the sensing site and removed by diffusion across the epitaxial layer and collection by the epitaxial junction, essentially all of the semiconductor area is available for photon charge generation. The response of the semiconductor is, of course, modified by the transmission characteristics of the surface conducting and insulating films. Charge generated in the neutral semiconductor is transported by diffusion and collected either at a sensing site or lost to the epitaxial collector. Sensing site spacing and epitaxial layer thickness consequently affect charge collection efficiency.

The absorption length of radiant energy into silicon ranges from $0.2\,\mu m$ at a wavelength of 400 nm to $5\,\mu m$ at 700 nm to $100\,\mu m$ at 1000 nm [2.7]. Good collection efficiency has been obtained with the epitaxial layer thickness roughly equal to the space between sensing sites.

Three conductor materials have been used in CID imager fabrication. The silicon gate process employs a lower level of polysilicon and an upper level aluminum conductor. Imagers have also been fabricated with two levels of polysilicon. A recent innovation [2.8] has been the incorporation of transparent metal oxide conductors into an overlapping electrode structure with polysilicon as the lower level conductor. The spectral transmission characteristics of these films (on quartz) is shown in Fig. 2.18. The range of responsivity that can be achieved using these techniques is shown in Fig. 2.19 for the sensing site layouts of Fig. 2.20. The $30\,\mu m$ square sensing site employs a double level polysilicon electrode structure on an epitaxial layer. The spacing between thin oxide regions is approximately equal to the undepleted epitaxial layer thickness.

The $35\,\mu m$ by $42\,\mu m$ site uses polysilicon for the lower conductor and metal oxide for the upper conductor. The spacing between thin oxide regions can be large on this array without loss in collection efficiency because it has been fabricated on bulk silicon.

2.3.2 Point Spread Function, Cross Talk, and Lag

The array cells are designed to allow charge transfer between the two storage capacitors at each sensing site but not between adjacent sites. Two mechanisms exist, however, which can result in signal cross talk between adjacent sites. Charge generated in the undepleted silicon between sensing sites will divide between the sites as long as the distance to be traveled is less than the minority carrier diffusion length. This effect can be attenuated in epitaxial imagers if the distance between the surface and the buried collector is made less than, or comparable to, the spacing between sites.

The second cross talk mechanism is the migration of injected charge to neighboring sites. Proper placement of the epi collecting junction also reduces this effect.

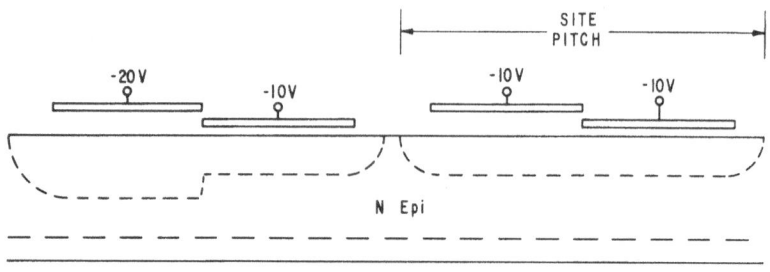

Fig. 2.17   P⁺ BULK

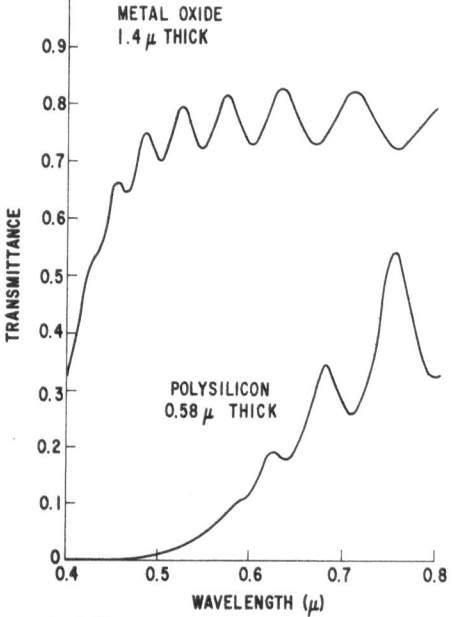

Fig. 2.18

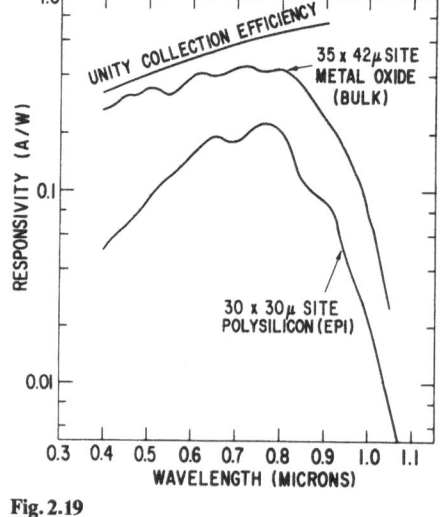

Fig. 2.19

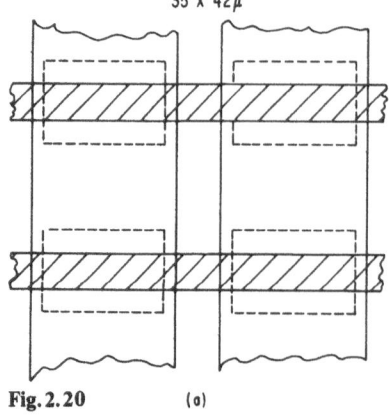

Fig. 2.20   (a)

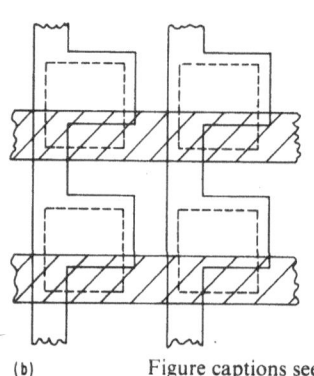

(b)   Figure captions see opposite page

Horizontal and vertical point spread functions for the 30 μm square sensing site of Fig. 2.20 is shown in Fig. 2.21. These measurements were made using a 2.5 μm diameter spot from a tungsten arc lamp.

Lag can occur in CID imagers through two mechanisms: 1) recapture of the injected charge by the injecting site upon reapplication of the storage electrode voltage; and 2) migration of injected charge to adjacent sites that are not to be read out until the following field. Image lag resulting from the second mechanism occurs to the same degree as cross talk (loss of MTF). Recapture of charge by the injecting site is a function of the time allowed for injection (see Fig. 2.6). This can vary from virtually zero recapture to full recapture of the injected charge (NDRO, for example). In practice, the injection pulse width is adjusted to ensure that image lag is consistent with system requirements.

2.3.3 Dark Current

The CID approach permits significantly more silicon area to be used for photon charge generation than for charge storage. This results in an advantageous dark current situation because the thermal charge generation rate in nondepleted bulk silicon is orders of magnitude less than in the depleted storage region [Ref. 2.1, pp. 173–180]. Consequently, each image sensing site collects and stores photon-generated charge from essentially the total site area but generates dark current only in the storage area. Also, no separate storage area is required for image readout, so that a dark current contribution from this source is avoided.

The use of bias charge in the storage area results in an additional reduction of dark current since the surface thermal generation rate in MOS structures is much smaller under inversion conditions than under depletion conditions [2.9]. Measured thermal charge generation rates for a 100×100 imager are shown in Fig. 2.22. Minimum dark current results from biasing the array such that charge is stored under both electrodes (Fig. 2.22a). Figure 2.22b shows the effect of operating with one storgage region depleted. These data were obtained by operating the device in a "burst" mode in which the sensing sites were allowed to integrate charge for the appropriate time after which the entire array was read out in 1/30 s. The average accumulated charge for the entire array was recorded.

The 100×100 imager is normally operated with one storage region depleted, so that the conditions of Fig. 2.22b apply. The peak signal-to-dark current ratio exceeds 100:1 under these conditions at 30 frames/s. For lower frame rates, it is feasible to operate under the conditions of Fig. 2.22a, thereby doubling the useful integration time.

◄ Fig. 2.17. Sensing site cross section

Fig. 2.18. Metal oxide and polysilicon film spectral transmittance

Fig. 2.19. Sensing site collection efficiency

Fig. 2.20a, b. Sensing site layouts (a) lower polysilicon row conductors, upper metal oxide conductors and (b) double level polysilicon conductors

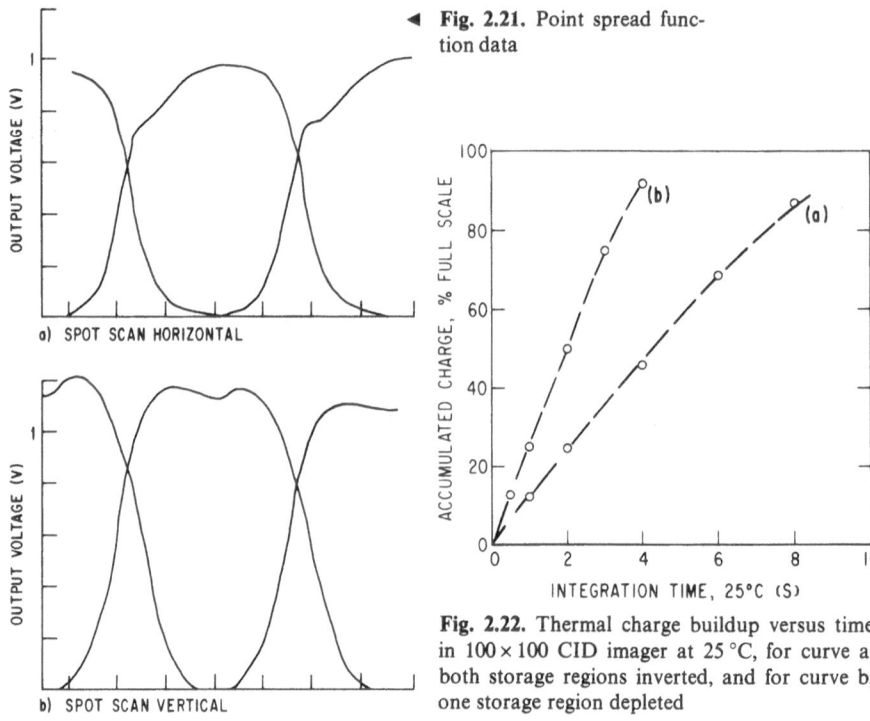

Fig. 2.21. Point spread function data

a) SPOT SCAN HORIZONTAL

b) SPOT SCAN VERTICAL

Fig. 2.22. Thermal charge buildup versus time in 100×100 CID imager at 25 °C, for curve a, both storage regions inverted, and for curve b, one storage region depleted

2.3.4 Blooming

The epitaxial CID structure is resistant to image blooming since each sensing site is electrically isolated from its neighbors. Charge spreading in the substrate is minimized by the underlying charge collector.

In sequential injection, blooming of the displayed image occurs if charge is injected from a half-selected site during readout of another site on the same column. This excess injected charge is detected as signal and adds to the displayed video. This results in brightening of the affected column upon overload of a single site.

The image displayed in the parallel injection approach exhibits relatively little blooming as a result of sensing site overload. This is because the half-select and injection operations occur during the horizontal blanking interval. While excess charge can accumulate during a line scan interval and cause column brightening for overloads occurring in the right-hand portion of the image field, this effect is attenuated by the line-to-frame integration time ratio.

During NDRO, virtually no blooming occurs, since the charge is not injected. The affected sites simply fill to capacity and cease collecting charge.

In all cases, radial spreading of excess charge is prevented by the underlying charge collector.

2.3.5 Noise Sources

The circuitry used to select and provide readout of CID image sensors contains a number of Johnson noise sources. The distributed resistance of the array lines used for signal sensing, the line selection switch, and the first preamplifier stage each contribute temporal noise to the video signal. In addition capacitor reset noise (kTC noise) can be significant when certain readout methods are used. Shot noise in the dark current and/or junction leakage current in the MOS line select multiplexers can be significant under certain conditions.

The Johnson noise sources usually are dominant in large arrays operating at high video rates.

If a switch is used to set the voltage across a capacitor, thermal noise in the resistance of the switch results in an uncertainty in the final capacitor voltage. The magnitude of this uncertainty [2.10] is

$$V_n = (kT/C)^{1/2} \tag{2.5}$$

or

$$Q_n = (kTC)^{1/2}, \tag{2.6}$$

where k Boltzmann's constant $= 1.38 \times 10^{-23}$ Ws/k, and T the temperature in Kelvin. Sequential injection CID imagers are not limited by this noise component because it is possible to reference the net injected charge signal to the input capacitor voltage after reset has been completed. This technique, called correlated double sampling [2.11], results in the substitution of kTC noise on a clamping capacitor for kTC noise on the array output capacitance. The level of kTC noise referred to the array can be made arbitrarily small, however, since gain can be used between the array output and the clamp capacitor. The parallel-injection technique does not allow complete elimination of kTC noise. The column reset transistors introduce kTC noise that is not rejected. Voltage noise at the input of the preamplifier results in an equivalent input charge that is directly proportional to the array output capacitance ($q = cv$). Theoretical preamplifier noise levels of a few hundred carriers result from array output capacitance levels in the 10 pF region. kTC noise can be either negligible or the predominant temporal noise source, depending upon the specific array design and readout method.

Under low video rate readout conditions, Johnson noise can be minimized by restricting the noise bandwidth of the video amplifier. Shot noise originating in array dark current and line select multiplexer junction leakage can be limiting under these conditions. Reduction in array operating temperature can be used to control these thermally generated currents and consequently the resultant shot noise.

Solid-state imaging sensors can exhibit a fixed non-uniform spatial background in the reproduced image. The major sources of fixed pattern noise in CID image sensors are transistor switching interference, array photolithographic variations, and bias charge variations. Nonuniform coupling of the MOS transistor scanner output voltage to the video signal results in a component of fixed pattern noise that repeats from scan to scan. Variations in row-to-column crossover capacitance arising from either insulator thickness or photolithographic variations cause a two-dimensional component of fixed pattern noise. Variations in bias charge from site to site, caused by differences in storage capacitance or threshold voltage, also result in a two-dimensional component of fixed pattern noise. Differential sensing and signal processing can be used to minimize these effects.

Dark current nonuniformity can be an important source of pattern noise, particularly at room temperature. The inherently low dark current performance of CID imagers can be an advantage under these conditions.

2.4 Special Features

The organization and nondestructive readout capability of the CID imager gives rise to a number of specialized functions. The $X-Y$ addressable organization allows random access to array pixels [2.12]. The nondestructive readout feature combined with $X-Y$ access allows linear combinations of pixel signals to be sensed giving rise to spatial transform readout [2.13]. Finally, signals can be repeatedly read and summed to improve dynamic range [2.14].

References

2.1   A.S.Grove: *Physics and Technology of Semiconductor Devices* (John Wiley and Sons, New York 1967)
2.2   P.G.Jespers, F.VandDeWiele, M.F.White: *Solid-State Imaging*, NATO Advanced Study Institutes Series E, No. 16 (Noordhoff, Leyden 1976) pp. 295–304
2.3   J.S.Brugler, P.G.A.Jespers: IEEE Trans ED-**16**, 297–302 (1969)
2.4   G.J.Michon, H.K.Burke: ISSCC Dig. Tech. Papers XVI, 138–139 (1973)
2.5   G.J.Michon, H.K.Burke: ISSCC Dig. Tech. Papers, Feb. 1974, pp. 26–27
2.6   G.J.Michon, H.K.Burke, D.M.Brown: "Recent Developments in CID Imaging", Proc. Symp. Charge-Coupled Device Technology for Scientific Imaging Applications (JPL) Vol. XVII, pp. 106–115 (1976)
2.7   S.M.Sze: *Physics of Semiconductor Devices* (Wiley Interscience, New York 1969) p. 661
2.8   D.M.Brown, M.Ghezzo, P.L.Sargent: IEEE Trans. ED-**25**, 79–84 (1978)
2.9   D.J.Fitzgerald, A.S.Grove: Surf. Sci. 347–369 (1968)
2.10 J.E.Carnes, W.F.Kosonocky: RCA Rev. **33**, 327–343 (1976)
2.11 M.H.White, D.R.Lampe, F.C.Blaha, I.A.Mack: IEEE J. SC-**9**, (**1**), 1–13 (1974)
2.12 H.K.Burke, G.J.Michon, T.L.Vogelsong: "Random Access CID Imager", Conf. on Laser and Electro-Optical Systems (CLEOS), Feb. 6–9, 1978, San Diego, CA
2.13 G.J.Michon, H.K.Burke, T.L.Vogelsong, P.A.Merola: "Charge Injection Device (CID) Hadamard Focal Plane Processor for Image Bandwidth Compression", Proc. Agard Symposium on Impact of CCD and SAW Devices on Signal Processing and Imagery in Advanced Systems, No. 230 (Oct. 11–15, 1977)
2.14 R.S.Aikens, C.R.Lynds, R.E.Nelson: "Astronomical Applications of Charge Injection Devices", in *Proc. Soc. Photo-Optical Instrum. Engrs., Low Light Level Devices*, Vol. 78 (Bellingham, WA 1976) pp. 65–71
2.15 D.F.Barbe: "Imaging Devices Using the Charge-Coupled Concepts", in Proc. IEEE **63**, 38–67 (1975)

3. Intrinsic Focal Plane Arrays

W. D. Baker

With 17 Figures

Focal plane arrays (FPAs) offer very attractive advantages for advanced infrared systems when compared to current technology. These advantages tend to fall into the categories of 1) higher performance, 2) system design flexibility, and 3) simplified systems. Higher performance is due to the very large number of detectors (and their high density) possible with focal plane arrays and to the use of time delay and integration (TDI) processing. The improved performance permits design trade-offs of such factors as aperture size and spectral bandwidth, resulting in design flexibility not otherwise possible. The elimination of low signal level leads through the dewar, the reduction in total dewar feedthroughs and the elimination of individual channel trimming results in system simplification.

3.1 Background

Several technology approaches for the development of ir FPAs have been proposed, and a number of these are being actively pursued. The various approaches differ in the implementation of the principal array functions: photon detection, detector readout, signal processing, and output multiplexing. Charge-coupled device (CCD) technology is a key element in all of the technology approaches. The major approaches include intrinsic detector arrays, extrinsic silicon arrays, and hybrid arrays. Intrinsic arrays are those in which the detection and readout are accomplished using an intrinsic semiconductor substrate material. Extrinsic silicon arrays – the topic of Chap. 4 – involve extrinsic detectors fabricated on the same substrate with conventional silicon readout devices. The hybrid array approach involves the interconnection of an array of detectors (usually intrinsic) to a separate silicon readout chip. As usual, each approach has its advantages and its problems. In this chapter, the characteristics of intrinsic focal plane arrays will be discussed, as well as some selected information on hybrid arrays.

Intrinsic FPAs have some definite strong points. These include:
1) small number of interconnects,
2) high quantum efficiency,
3) higher operating temperature than extrinsic silicon,
4) potential for selectable spectral response with some materials.

A major disadvantage of intrinsic arrays is the relatively primitive state of the technology for fabricating FPAs on useful intrinsic materials. In addition, there may also be fundamental limitations on the device performance of some of the intrinsics as a result of specific physical properties.

Hybrid FPAs arise from an attempt to combine the most highly developed CCD and signal processing materials, silicon, with the generally desirable properties of intrinsic detectors. This approach allows great flexibility in choice of detector characteristics and may allow for generic silicon devices capable of being used with a variety of detectors. The problem areas in the hybrid approach are the mechanical interface (with the potential for thermal mismatch, bad contacts, or difficult alignment) and the electrical interface between the detector and the CCD. In most cases these arrays require one interconnect per detector, although with charge-injection device (CID) detectors this can be reduced to one interconnect for each row and each column.

3.2 Device Configurations

3.2.1 Approaches

Three possible configurations for intrinsic FPA are shown schematically in Fig. 3.1. These include: a) a charge-injection device detector array in conjunction with a separate signal processor; b) a charge-coupled device array in which the signal charge is collected and integrated in the shift register sites; and c) a CCD array in which separate photosites are used for signal charge collection and integration prior to charge transfer into a shift register. In both a) and b), the actual detector is an metal-insulator semiconductor (MIS) element. For this detector, photons entering the semiconductor (through semitransparent gates for frontside illumination or through either a transparent substrate or a thinned substrate for backside illumination) generate signal charge which is collected by a depletion region under the gate. Except for the use of a semiconductor with a band gap suitable for the ir, this detection process is identical to that for visible CID and frame transfer CCD imaging. In case c), the photosite may be an MIS detector or it may be a photovoltaic (PV) detector. This configuration then basically operates the same way as an interline transfer visible CCD imager or as a visible photodiode array with CCD readout.

In principle, all of these array configurations can be operated in a staring mode. That is, of course, the mode in which visible imagers are usually operated and for a review of which see *Barbe* [3.1, 2]. For tactical ir applications involving terrestrial backgrounds, however, the responsivity variations and offsets across a reasonable size device currently are simply too large to permit staring operation. Additionally, the typical background levels are not compatible with integration for full TV frame times with reasonable size CCD wells.

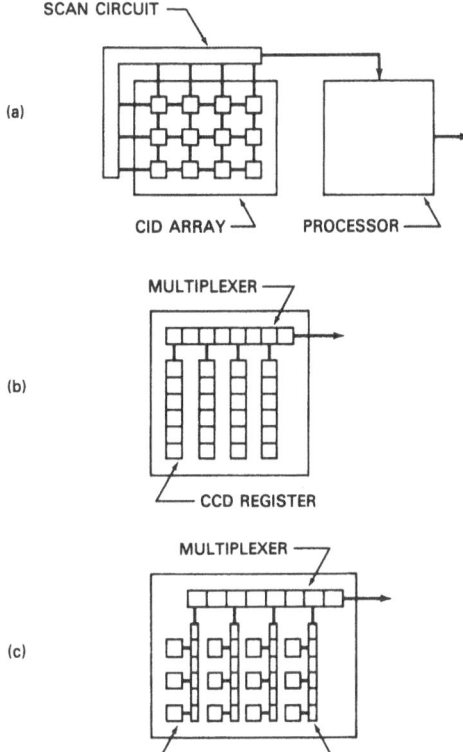

(a)

(b)

(c)

SCAN CIRCUIT

CID ARRAY   PROCESSOR

MULTIPLEXER

CCD REGISTER

MULTIPLEXER

PHOTOSITES   CCD REGISTER

Fig. 3.1a–c. Intrinsic focal plane array configurations

Efforts are underway to reduce variations across arrays, to develop simple signal processing for uniformity correction, and to develop techniques for background signal suppression. If these efforts are successful, then staring operation of ir FPAs will be practical. Until that time, most tactical applications are expected to use scanned focal planes with TDI signal processing. For the CID configuration of Fig. 3.1 the TDI would be performed on a separate processor chip, probably using silicon technology. In the other two configurations, the CCD shift register would provide the TDI processing.

All of the proposed configurations for intrinsic FPAs depend heavily on the availability of a MIS technology for intrinsic ir materials.

3.2.2 Charge-Injection Devices

The CID FPA (shown in Fig. 3.1a) is an intermediate technology between most hybrid arrays (with separate detectors and one interconnect per detector) and monolithic arrays which include at least some of the signal processing on the detector substrate. A CID structure has the virtue that it can be fabricated in *materials which* have an interface state density which is too large for CCD

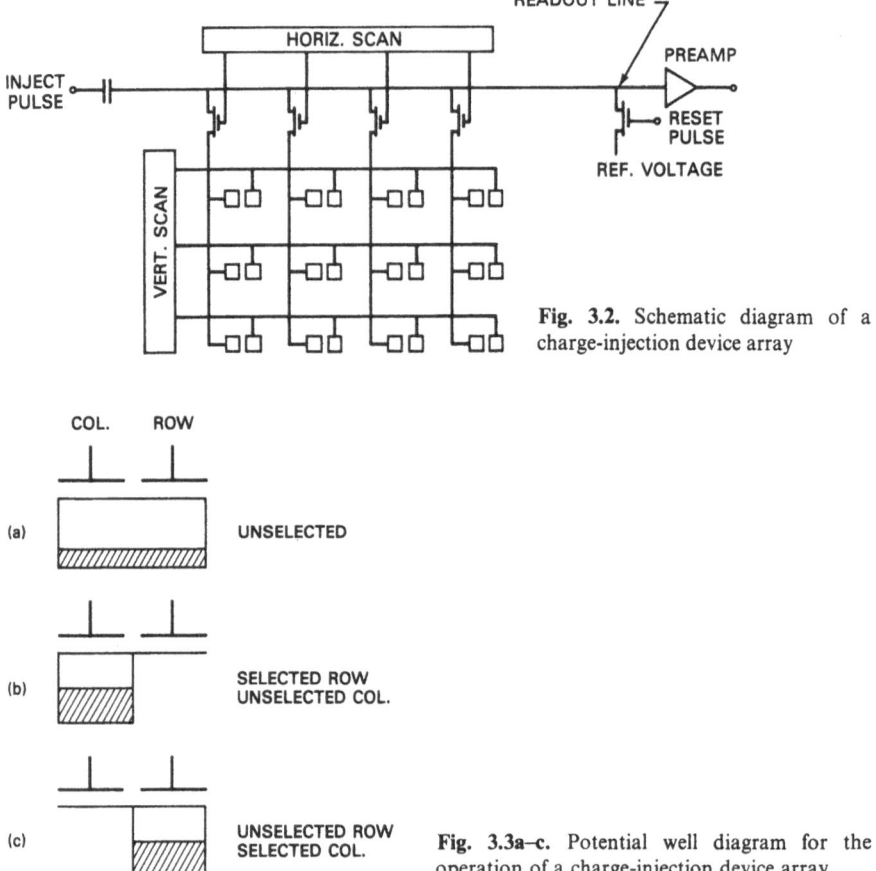

Fig. 3.2. Schematic diagram of a charge-injection device array

Fig. 3.3a–c. Potential well diagram for the operation of a charge-injection device array

operation. In addition, the CID array can be randomly addressed, which could be of significance for some staring applications, and can be relatively insensitive to blooming [3.3].

The form of a CID array is shown in Fig. 3.2. In order to permit $x - y$ coincidence addressing, each photosite is divided into separate row and column electrodes. The essentials of the device operation are illustrated in Fig. 3.3 for one photosite. Figure 3.3a indicates the situation when neither the row nor the column site has been addressed and some charge has collected in the wells. In order to read the array, one row is first selected and the potential wells connected to that row are collapsed. Any charge residing under the row electrodes in the selected row is transferred to the potential well under the adjacent column electrode. Figure 3.3b shows this condition. After the row is selected, each column is sequentially addressed and the potential wells connected to that column are collapsed. At photosites which are not located in the selected row, the charge that was under the column electrode flows to the

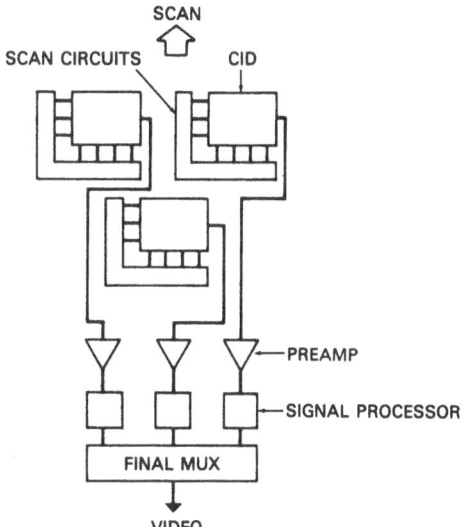

Fig. 3.4. A possible focal plane assembly using charge-injection device array modules

potential well under the row electrode, as shown in Fig. 3.3c. Only at the photosite at the intersection of the selected row and the selected column are both potential wells collapsed, resulting in the injection of the charge into the substrate. The signal charge at each site could be readout by measuring the substrate current resulting from the charge injection. In practice, however, other read-out schemes work better. One such readout scheme is sequential injection [3.3] which is suitable for data rates up to about 2 MHz. The readout sequence is as follows (refer to Fig. 3.2): the horizontal scan register is operated to connect one column to the readout line, which is floating; then the row scan register is used to select a row, causing the transfer of charge to the column electrodes along that row; next the readout line and the connected column are reset to a reference voltage. At this point, an injection pulse is ac coupled to the readout line and injects the charge at the coincident photosite. The change in the readout line potential caused by the injection of the charge is sensed by the output preamplifier. After the readout is repeated for all of the columns in the selected row, the next row is selected and the process repeated.

The processing technology available for a given intrinsic material may not be adequate for the fabrication of the scan registers, selection switches, or preamplifiers. In such a case, the CID array itself is fabricated in the intrinsic material and the other elements, plus TDI registers or other signal processing devices, are fabricated using conventional silicon technology. A scanned focal assembled from modules of this type is shown in Fig. 3.4. The system characteristics of such a focal plane have been analyzed [3.4a]. In this arrangement, the connections required are one for each row, one for each column, and one signal connection for every CID module. Some of the signal processing shown in Fig. 3.4 may be done off the focal plane, and perhaps outside the dewar, depending on the particular system. One problem with this approach is the relatively high data rate required from the CID array. Since the TDI function is performed on a separate device, and the detector outputs are effectively premultiplexed, the entire CID array must be readout and the signals loaded into the TDI register in one dwell time (or less, if more than one sample per dwell is required). For some system designs, this data rate may result in an unrealistic bandwidth for the preamplifiers. In such cases, it may be desirable to use more than one preamplifier per CID module, with the limiting case being one preamplifier per column [3.4b].

Another potential problem with the CID approach is the relatively high capacitance of the signal line from the array. This line has the capacitance of the readout line plus one entire column of photosites and strays. The high capacitance results in a low noise voltage for a given noise charge on a photosite since

$$V_n = Q_n/C_n .$$

As a result, the preamplifier must have a very low equivalent input noise voltage if the total system noise is to be dominated by the background noise. Such a low noise when combined with the required bandwidth can be a very severe requirement on the preamplifier.

3.2.3 Charge-Coupled Devices

The CCD FPA shown in Fig. 3.1b utilizes the shift register sites themselves as the photosites. The focal plane filling possible with this approach is very attractive, and the fabrication process is potentially simple. For scanned arrays, the CCD register performs the TDI function. Since the TDI is performed on the detector substrate without any premultiplexing, the data rates are lower by a factor equal to the number of detectors in the TDI direction than is the case for the CID FPA. One potential problem with this approach is the signal handling capacity of the CCD register, especially for high background applications, since the background will be added at each TDI step. In the CID approaches discussed previously, the signal can be ac coupled prior to the TDI operation.

Although Fig. 3.1b shows the multiplexer register on the same substrate as the detector-TDI register, the fabrication technology for a given intrinsic material may not allow this. The multiplexer must run m times faster than the TDI register, where m is the number of columns. High-speed CCDs require much better technology in order to obtain adequate transfer efficiency. If the technology is limited, the multiplexer may be fabricated on a separate chip using conventional silicon technology, and the signal brought off the array at the end of each TDI channel. The signal at that point, however, is a higher level signal with a good signal-to-noise ratio as a result of the TDI. In addition, the output of the CCD can be a low capacitance node. Consequently, the only drawback to such an approach is the increased number of connections required to the array.

The array configuration in Fig. 3.1c has separate detector sites connected to the CCD registers, which in scanned systems perform the TDI function. This structure is more complicated than that of Fig. 3.1b since some interface is required between the detector site and the CCD. In addition, it is harder to achieve large detector densities since each detector column must have an associated CCD column. The opportunity to do ac coupling, background suppression, or other functions between the detector and the TDI register can solve the dynamic range problem for the CCD since it does not need to hold the full integrated background signal charge. The output multiplexer shown in Fig. 3.1c may be located on a separate silicon device, as discussed previously.

A focal plane using either type of CCD array would probably be simpler to assemble than one using a CID array. The question that remains is when an MIS technology suitable for CCD fabrication in ir intrinsics will be developed.

3.3 Device Analysis

3.3.1 Materials Requirements

Both CID and CCD structures have certain device characteristics which directly affect the application of those structures to ir FPAs. Those device characteristics result from the fundamental properties of the particular semiconductor, from the properties of the MIS interface, and from the design of the structure. Some of the important device characteristics and the basic properties that control those characteristics are shown in Table 3.1. Semiconductors being considered for use in ir FPAs must be evaluated on the basis of their physical properties and on the quality of the MIS technology available or projected for that semiconductor.

The two spectral bands of interest for most tactical ir imaging or detection systems are the 3–5 μm and the 8–12 μm wavelength atmospheric windows. The principal semiconductor families whose bandgaps provide cutoff wavelengths for one of these spectral bands are:

3–5 μm	8–12 μm
InSb	HgCdTe
InAsSb	PbSnTe
InGaSb	
HgCdTe	

The basic requirements for a semiconductor for use in MIS devices are 1) a reasonable dielectric constant, 2) long lifetime, and 3) relatively low substrate *doping. Of the 3–5 μm materials* listed above, InSb, alloys from the

Table 3.1. Some important material or device properties and the device characteristics which they affect

Device type	Device characteristic	Basic properties										
		α	E_g	n_i	τ	S	L	N	W	N_{fs}	C_0	V
CCD and CID	Cutoff wavelength		×									
	Quantum efficiency	×				×	×	×				
	Cross talk	×					×					
	Dynamic range										×	×
	Dark current	×	×	×	×	×	×	×	×			×
CID	Injection time				×							
CCD	Transfer efficiency									×		

α: optical absorption coefficient
E_g: energy gap
n_i: intrinsic carrier density
τ: lifetime
S: surface recombination velocity
L: diffusion length

N: substrate doping density
W: depletion width
N_{fs}: interface state density
C_0: insulator capacitance
V: gate voltage

In-Ga-As-Sb family, and HgCdTe all have promising characteristics. The Pb-salts family, however, suffers from a very high dielectric constant, relatively high substrate doping, and relatively short lifetime. These features combine to make MIS devices, and FPAs based on them, unlikely in this material. Indium antimonide is a well-developed material, but has a fixed energy gap (and so a fixed cutoff wavelength). The alloys of the In-Ga-As-Sb family are much less well developed, but do offer a selectable cutoff wavelength by composition control. The variation of energy gap with composition is shown for the InAsSb and InGaSb systems in Fig. 3.5. One problem with this material system is a relatively poor lattice constant match between the alloys with a useful bandgap and the GaSb or InSb bulk material which might be used as substrates for epitaxial growth. Figure 3.6 illustrates the magnitude of this problem. The HgCdTe system also offers a selectable cutoff wavelength, as shown in Fig. 3.7, and the growth of bulk crystals (in small sizes) is reasonably well developed. Larger substrates of CdTe are available for epitaxial growth with only a modest lattice mismatch. In the 8–12 μm band, the only presently attractive material for intrinsic FPAs is HgCdTe.

Quantum efficiency depends on the losses which may occur in getting photons into the semiconductor, on the absorption of the photons, and on the collection of the photon-generated signal charge by the potential wells.

Cross talk is the introduction of signal charge belonging in one detector into the signal charge packet of another detector. Optical cross talk may result if the photons are absorbed outside of the detector potential well so that the resulting carriers can diffuse to an adjacent detector site, or if the photons are not fully absorbed in the semiconductor on the first pass and are then reflected

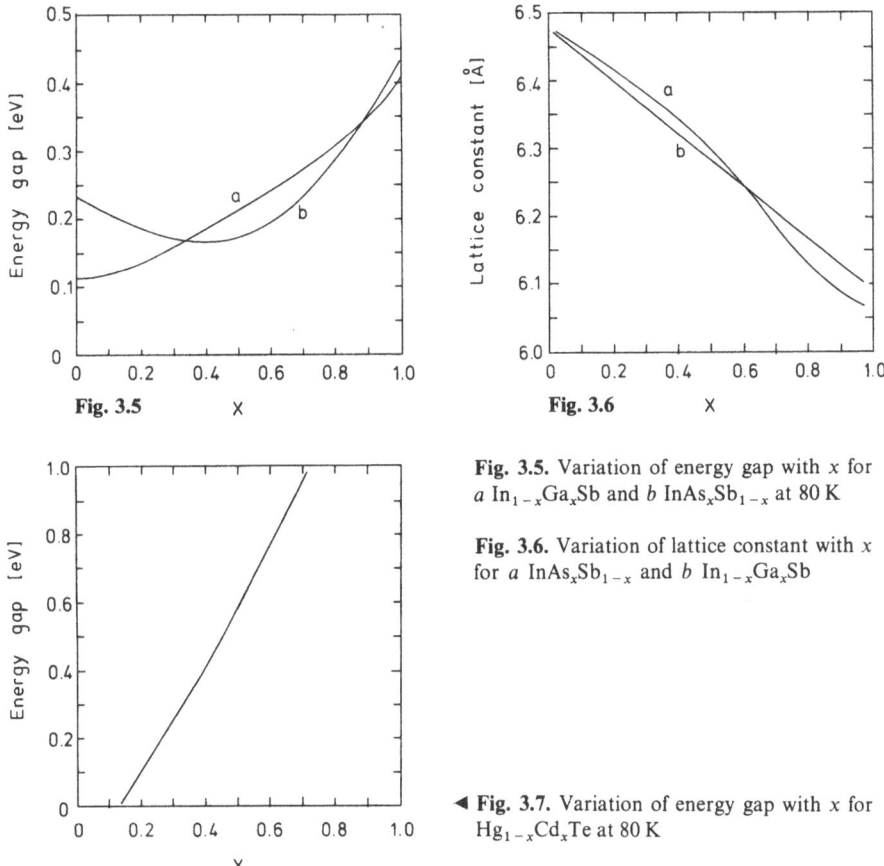

Fig. 3.5

Fig. 3.6

Fig. 3.5. Variation of energy gap with x for a $In_{1-x}Ga_xSb$ and b $InAs_xSb_{1-x}$ at 80 K

Fig. 3.6. Variation of lattice constant with x for a $InAs_xSb_{1-x}$ and b $In_{1-x}Ga_xSb$

◀ **Fig. 3.7.** Variation of energy gap with x for $Hg_{1-x}Cd_xTe$ at 80 K

from the back interface. This latter situation is unikely for intrinsic materials unless they are very thin. Electrical cross talk can result from charge transfer inefficiency in the CCD TDI or multiplexer registers. In this case, charge from one packet is trapped and later released into a following packet. Electrical cross talk may also arise from capacitive coupling between interconnect runs on the array, for example if clock lines are allowed to overlap low level signal lines.

Dynamic range depends on the maximum signal capacity of the detector site, the interface to the CCD, the CCD registers, or the subsequent signal processing electronics – whichever is less – and the array noise which determines the minimum detectable signal. The desired situation is enough well capacity to tolerate the largest signal possible and a noise level dominated by the shot noise on the background signal.

Array dark current limits the maximum operating temperature of the intrinsic FPA. Dark currents generally arise from generation at the device surface, generation in the neutral substrate volume, and generation within the depletion region. These dark current components are controlled by processing to obtain long bulk lifetimes and low surface state densities and by selection of the operating temperature. Another potential source of dark current in CCD or CID structures is band-to-band tunneling which can occur when band bending takes place at the insulator-semiconductor interface.

In the operation of a CID, the charge injected at a site must have time to recombine before the potential well can be reestablished. The lifetime of the excess injected charge must be short compared to the time between the injection of the charge and the readout of the site. If this lifetime is not sufficiently short in the bulk material, a buried junction can be used to collect the injected charge [3.3].

The transfer efficiency of a CCD can depend on the geometry of the device, on the clocking waveforms, and on interface or bulk trapping mechanisms. In the case of the infrared materials being considered, surface channel devices are the most likely structures due to the relatively limited fabrication technology available. In surface channel CCDs, a bias charge (or fat zero) can be maintained in the device to keep the interface traps nearly filled. If that is done, the transfer speed is limited by free charge transfer from one gate to the next under the influence of self-induced drift, diffusion, and fringing-field drift. The interface state trapping also introduces a noise component which can be significant in some array applications.

3.3.2 Quantum Efficiency

If the arrays are illuminated from the front (the electrode) side, a detector site gate material which is reasonably transparent to the wavelengths of interest must be used. Some work using very thin (50–100 Å) metal films such as titanium [3.5] and nichrome [3.6] has been reported, but more effort is required in this area. For frontside illumination, the gate insulator must also be chosen for minimal absorption at the wavelengths of interest. One alternative to frontside illumination is to thin the substrate to a thickness on the order of a diffusion length and then to illuminate from the back. This thinning introduces processing complexities, usually including the need to accumulate the back-side surface, but it eliminates absorption losses in the gate metal and insulator. A second backside illumination alternative is to grow an epitaxial layer of the intrinsic material on a wider band gap substrate. The array is then illuminated from the back, through the transparent substrate. A backside-illuminated approach is possible for the device configurations shown in Fig. 3.1a and b, but is generally impractical for the configuration of Fig. 3.1c due to the difficulty of shielding the CCD registers from the illumination. Of course, any reflectance losses at the various interfaces can limit the overall quantum efficiency of an array and so may need to be considered during device design.

Assuming that any absorption in the electrode or insulator materials has already been considered, and that internal reflection in the semiconductor is neglibible, then the quantum efficiency of the FPA is determined by the efficiency of the collection of generated carriers. For uniform illumination, the quantum efficiency η is given by [3.7]:

$$\eta = \left[1 - \frac{\exp(-\alpha w)}{(1 + \alpha L)} \right] T, \tag{3.1}$$

where α is the optical absorption coefficient, w is the depletion region width, L is the minority carrier diffusion length, and T includes all reflectance and transmission effects. With respect to the semiconductor itself, the quantum efficiency will be nearly unity as long as the second term inside the brackets is small. For intrinsic materials like InSb or HgCdTe, the absorption coefficient is large and the diffusion length is reasonably long, so that this condition is generally fairly well satisfied. To minimize cross talk and to ensure that the quantum efficiency is constant with time, it is desirable that almost all of the photon absorption take place within the depletion region of the detector site. This will occur if the term $\exp(-\alpha w)$ is much less than unity. For InSb and HgCdTe, α is about $5 \times 10^3 \, \text{cm}^{-1}$ or greater (see, for example, [3.8]). If the αw product is to be 2 or 3, then the depletion width must be 4–6 μm. Depletion widths in this general range are probably possible, at least in some materials.

3.3.3 Basic MIS Analysis

A simple MIS cell and the corresponding energy band diagram are shown in Fig. 3.8. The cell shown assumes an n-type semiconductor substrate (or p-channel technology). Since p-channel MIS technology is somewhat easier to do than n-channel, most of the current development work in intrinsic FPAs has been directed toward p-channel devices. In operation, a voltage is applied to the MIS gate in order to establish a depletion region (a potential well). Initially, no charge is present in the potential well and the diagram of Fig. 3.8a applies. At a detector site, incoming photons then generate a signal charge which is collected in the potential well. Since the collected charge seeks the minimum energy, it is stored very near the insulator-semiconductor interface, as shown in Fig. 3.8b. Because of the direction of the energy band bending, the carriers collected by the well are minority carriers in the semiconductor and they represent an inversion charge at the interface. The inversion charge effectively reduces the applied voltage to the semiconductor, reducing the depletion width. Note that the capacitance of the device is the series combination of the insulator capacitance and the depletion region capacitance. In the limit, when enough charge is collected the depletion region will be reduced to zero width and no further charge can be collected at the detector. At this point the device capacitance is the insulator capacitance.

An analysis of the MIS device basically requires the solution of Poisson's equation for the system. Summaries of the approach are given by *Grove* [3.9] and *Sze* [3.10]. The result of the analysis is that for most devices of interest the

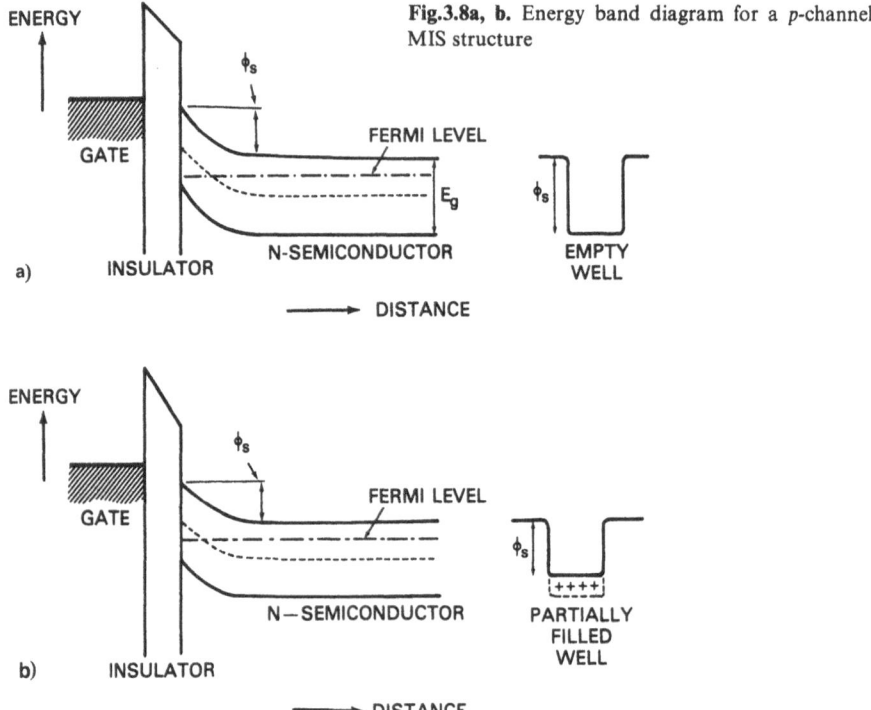

Fig.3.8a, b. Energy band diagram for a *p*-channel MIS structure

surface potential ϕ_s is given by the depletion approximation which for a *p*-channel devices gives

$$\phi_s = -qN_Dw^2/2\varepsilon_s\varepsilon_0 \tag{3.2}$$

where q is the magnitude of the electronic charge, N_D is the substrate doping density (*n*-type), ε_s is the dielectric constant of the semiconductor, ε_0 is the permittivity of free space, and w is the depletion width.

For an example, select InSb as a potentially useful material for an FPA. The dielectric constant for this semiconductor is 17. In MIS devices for FPAs, low substrate doping is generally desirable in order to obtain long lifetimes, high breakdown voltages, low tunneling currents, and wide depletion widths for a given gate voltage. The range of minimum substrate doping available for InSb (and for HgCdTe as well) of good quality is typically from 1×10^{14} to 1×10^{15} cm^{-3}. A number of deposited insulators are potentially useful for MIS devices on ir materials. One group of these – SiO$_2$, SiO, and some SiO$_x$N$_y$ – have dielectric constants of approximately 4. A second group – Si$_3$N$_4$ and ZnS – have dielectric constants of approximately 7. For example, SiO$_x$N$_y$ has been applied to InSb devices [3.6] and ZnS has been applied to HgCdTe [3.11]. Other insulators are possible, including native insulators formed on HgCdTe

[3.12]. The insulator thickness is restricted by device and fabrication considerations. If the insulator is too thin, pinholes occur which result in poor device yield. If the insulator is too thick, the applied gate voltage necessary to obtain an acceptable potential well becomes too large. With deposited insulators, the minimum practical thickness for acceptable yield is in the range of 1000 Å. For reasons which will be discussed later, the surface potential must be limited to a relatively low value, say a few volts, for the ir intrinsic semiconductors. This means that the effective gate voltage V_G is·limited to a few volts – perhaps 1 to 5. Since a typical flat-band voltage for an MIS structure on InSb or HgCdTe is roughly 2 to 5 V, the applied gate voltage would be in the range of 3 to 10 V for a typical device.

By inspection of Fig. 3.8,

$$V_G = \phi_s + V_{ins},\tag{3.3}$$

where V_G is actually the effective gate voltage, which is the applied gate voltage reduced by the flat-band voltage of the MIS structure, and V_{ins} is the voltage across the insulator. Applying Gauss' law, we can write

$$V_{ins} = -(Q_i + Q_B)/C_0,\tag{3.4}$$

where Q_i is the charge per unit area in the inversion layer, Q_B is the charge per unit area in the depletion region, and C_0 is the insulator capacitance per unit area. Note that $C_0 = \varepsilon_0 \varepsilon_i/t$ where ε_i is the dielectric constant and t is the thickness of the insulator. For the situation of Fig. 3.8,

$$V_{ins} = -(Q_i + qN_D w)/C_0.\tag{3.5}$$

Combining (3.2), (3.3), and (3.5) to eliminate w gives

$$V_G = \phi_s - \frac{Q_i}{C_0} - \frac{qN_D}{C_0}\left(\frac{-2\varepsilon_s \varepsilon_0 \phi_s}{qN_D}\right)^{1/2}.\tag{3.6}$$

This equation can be solved for the surface potential,

$$\phi_s = (V_G + Q_i/C_0) - qN_D \varepsilon_0 \varepsilon_s/C_0^2$$
$$+ [-2q\varepsilon_0 \varepsilon_s N_D(V_G + Q_i/C_0) + (qN_D \varepsilon_0 \varepsilon_s/C_0)^2]^{1/2}/C_0.\tag{3.7}$$

For the cases of interest in a *p*-channel device, V_G is negative and the inversion charge is positive. For a given set of conditions, (3.7) can be solved for the surface potential and then (3.2) can be solved for the width of the depletion region. As an illustration, the calculation of surface potential and depletion width will be done for some specific cases.

Table 3.2. Example results for an InSb–SiO$_2$ device

Given			Calculated	
V_G [V]	Q_i [C/cm^2]	N_D [cm^{-3}]	w [µm]	ϕ_s [V]
-2	0	1×10^{15}	1.6	-1.3
-3	0		2.0	-2.1
-4	0		2.3	-2.9
-2	0	1×10^{14}	5.7	-1.7
-3	0		7.1	-2.7
-4	0		8.2	-3.6
-2	2.2×10^{11}	1×10^{14}	3.9	-0.8
-3	3.3×10^{11}		4.8	-1.2
-4	4.4×10^{11}		5.7	-1.7

The maximum charge (representing signal charge, background charge, and dark current) which can be stored in a potential well is called the well capacity. In the limit, charge can be collected until the depletion region is completely collapsed. At that point

$$Q_i = C_0 V_G. \tag{3.8}$$

In practice, the well capacity is usually limited by the barrier heights between adjacent potential wells and is generally in the range of half the value predicted from (3.8).

Based on the considerations just discussed, consider the following MIS device:

InSb substrate, n-type

SiO$_2$ insulator, 1000 Å thick.

For this device, we can calculate $C_0 = 3.5 \times 10^{-8}$ F/cm^2 and then construct Table 3.2 from (3.7) and (3.2) for the conditions of an empty well and a full well [taken as half the value given by (3.8)]. From Table 3.2, it can be seen that in order to achieve depletion widths in the range of a 4 µm (as needed for good quantum efficiency and low-optical cross talk) while keeping the surface potential to a few volts, doping densities on the order of 1×10^{14} cm^{-3} are required. For reference, a well capacity of 2.2×10^{11} charges/cm^2 corresponds to about 1.5×10^6 carriers for a 25×25 µm^2 device.

If a higher dielectric constant insulator such as Si$_3$N$_4$ or ZnS is used, a device design trade-off compared to the example just shown it possible. One choice may be to increase the insulator thickness to improve processing yield, while keeping C_0 constant. In the case of Si$_3$N$_4$, the insulator thickness could be increased to about 1800 Å from the 1000 Å for SiO$_2$ and all of the results of Table 3.2. would still be valid. Another choice would be to keep the insulator thickness at 1000 Å and take an increase in C_0. This choice results in a larger well capacity and in an increased depletion width for a given effective gate voltage.

In the case of a detector site, the inversion charge accumulates as a result of photon generation and as a result of device dark current. The charge due to photon generation is given by

$$dQ_i/dt = q\eta\Phi,\tag{3.9}$$

where Φ is the photon flux density arriving at the detector site. If the device and material properties result in a constant quantum efficiency in time, then

$$Q_i = q\eta\Phi T_i,\tag{3.10}$$

where T_i is the integration time for the detector.

3.3.4 Dark Current

Dark current in an MIS device results in the collection of a dark current charge in the potential wells of the device. This charge can limit the signal handling capacity (dynamic range) of the device and results in a noise component due to the variance on the charge. The sources of dark current include a) generation in the semiconductor or at the insulator-semiconductor interface, b) band-to-band tunneling, c) avalanche generation at channel stop or similar diffusion boundaries, and d) point defects which produce localized dark current "spikes."

Generation current arises from three mechanisms: 1) generation from surface states at the insulator-semiconductor interface, 2) generation in the depletion region of the device, and 3) generation in the neutral region of the semiconductor with diffusion of the carriers to a potential well. The generation current depends on the detailed parameters of the states involved, and on the way in which the device is operated. To estimate the magnitude of the current for intrinsic devices, the usual first order theory will be applied.

The current density due to generation from surface states can be written as [3.9]:

$$J_{gs} = qs(n_s p_s - n_i^2)/(n_s + p_s + 2n_i)\tag{3.11}$$

for the case of a single energy state located at midgap with equal hole and electron capture cross sections and where s is the surface generation velocity, n_i is the intrinsic carrier density, n_s is the surface electron density, and p_s is the surface hole density.

For a completely depleted surface, $n_s = p_s = 0$ and the current density is

$$J_{gs} = qsn_i/2. \tag{3.12}$$

If, however, a significant inversion charge is maintained in the potential well, the resultant screening of the surface leads to a greatly reduced generation current,

$$J_{gs} = qn_i sn_i/(p_s + 2n_i) \tag{3.13}$$

for an n-type substrate. This is generally written as

$$J_{gs} = qn_i s_e, \tag{3.14}$$

where s_e is the effective surface generation velocity, and is a function of the surface potential. For this reason, a bias charge is generally maintained in MIS detector sites to suppress the dark current generation. For a bias charge of a few percent of the maximum well capacity, values of $s_e < 20$ cm/s are normally obtained in InSb, based on a Zerbst analysis [3.5]. With such a bias charge, the surface generation current is reduced by a few orders of magnitude from the case of a totally depleted surface. The intrinsic carrier density depends on the particular semiconductor material and on the temperature. For InSb the intrinsic carrier density can be approximated from [3.13]

$$n_i = 5.7 \times 10^{14}(T)^{1.5} \exp(-0.125/kT), \tag{3.15}$$

where T is the temperature (Kelvin) and kT is the thermal voltage (in eV). In the case of $Hg_{1-x}Cd_xTe$, the intrinsic carrier concentration is given by [3.14]

$$n_i = (1.093 - 0.296x + 0.000442\ T)\ 4.29 \times 10^{14} E_g^{0.75} T^{1.5} \exp(-E_g/2kT), \tag{3.16}$$

where E_g is the energy gap given by [3.15]

$$E_g\,[eV] = -0.25 + 1.59x + (5.23 \times 10^{-4})(1 - 2.08x)\,T + 0.327x^3. \tag{3.17}$$

At 77 K, for InSb (with an energy gap of 0.24 eV) $n_i \sim 3 \times 10^9$ cm^{-3}. For $Hg_{1-x}Cd_xTe$ at 77 K, if $E_g = 0.25$ eV $(x \sim 0.3)$ then $n_i \sim 1 \times 10^9$ cm^{-3} and if $E_g = 0.10$ eV $(x \sim 0.21)$ then $n_i \sim 3 \times 10^{13}$ cm^{-3}. Assuming an effective surface generation velocity of about 10 cm/s, the surface generation current for either InSb or 0.25 eV HgCdTe is in the range of 2 to 5 nA/cm^2 while for 0.1 eV HgCdTe it is in the range of 50 µA/cm^2.

The current density due to generation in the neutral bulk region can be modeled for an n-type substrate by

$$J_{gb} = qn_i^2 L_p/N_D \tau_p, \tag{3.18}$$

where L_p is the hole diffusion length and τ_p is the hole lifetime. The diffusion length can be estimated from $L_p = (D_p \tau_p)^{1/2}$ where $D_p = kT\mu_p/q$ with μ_p the bulk hole mobility. The value of the lifetime and the mobility depend quite strongly on the quality of the semiconductor material. For InSb, mobility values of about 1×10^4 cm^2/V s and lifetimes of 0.1 to 1 μs have been reported [3.5]. For HgCdTe, mobility values of 600 cm^2/V s and lifetimes about 5 μs have been reported at 77 K [3.16]. Assuming a substrate doping density of 1×10^{14} cm^{-3} leads to current densities of less than 1 nA/cm^2 for both InSb and 0.25 eV HgCdTe at 77 K, but a current density of about 1 mA/cm^2 for 0.10 eV HgCdTe at 77 K. Reducing the temperature to 60 K would reduce the current density for the 0.10 eV HgCdTe by more than an order of magnitude.

The current density due to generation in the depletion region can be written as

$$J_{gd} = q n_i w / \tau_g, \tag{3.19}$$

where w is the width of the depletion region and τ_g is the generation lifetime. If the generation in the depletion region is dominated by a simple generation-recombination center, the generation lifetime can differ dramatically from the minority carrier lifetime, depending on the position of the center in the energy gap of the semiconductor [3.17]. In particular, for a single level center modeled by Shockley-Read-Hall statistics,

$$\tau_g = \tau_p \exp(E_t/kT) + \tau_n \exp(-E_t/kT), \tag{3.20}$$

where E_t is the position of the trap with respect to midgap and τ_n, τ_p are the minority carrier lifetimes. In the special case of a center at midgap with equal hole and electron lifetimes, $\tau_g = \tau_p/2$. For centers a few kT in energy away from midgap, the generation lifetime can be significantly larger or smaller than the minority carrier lifetime. Measurements on InSb suggest generation lifetime values in the range of 0.1 to 1 μs [3.5, 18] which is reasonably consistent with a midgap center. For HgCdTe, a midgap center would result in a generation lifetime in the range of 1 to 10 μs. If the depletion region width is about 5 μm (corresponding to a surface potential of approximately 1.5 V for a substrate with 1×10^{14} cm^{-3} doping), the resulting current density is in the vicinity of 0.2–2 μA/cm^2 for InSb, 0.01–0.1 μA/cm^2 for 0.25 eV HgCdTe, and 250–2500 μA/cm^2 for 0.10 eV HgCdTe at 77 K.

A summary of the generation current estimates is given in Table 3.3. Since the calculations depend on the values of the material parameters, any of the components could vary substantially from those shown in the table if a different set of parameters is assumed or measured. It should also be noted that in general both the generation in the depletion region and the generation at the interface depend on the surface potential and so are really functions of time for an MIS device.

Table 3.3. Summary of generation current calculations

	InSb	HgCdTe	
E_g	0.24 eV	0.25 eV	0.10 eV
n_i	3×10^9 cm^{-3}	1×10^9 cm^{-3}	3×10^{13} cm^{-3}
τ_p	0.1 µs	5 µs	5 µs
μ_p	1×10^4 cm^2/Vs	600 cm^2/Vs	600 cm^2/Vs
s_e	10 cm/s	10 cm/s	10 cm/s
τ_g	0.2 µs	10 µs	10 µs
J_{gs}	5 nA/cm^2	2 nA/cm^2	50 µA/cm^2
J_{gb}	0.4 nA/cm^2	1 pA/cm^2	1 mA/cm^2
J_{gd}	1 µA/cm^2	80 nA/cm^2	240 µA/cm^2
	$T = 77$ K, $N_D = 1 \times 10^{14}$ cm^{-3}, $w = 5$ µm		

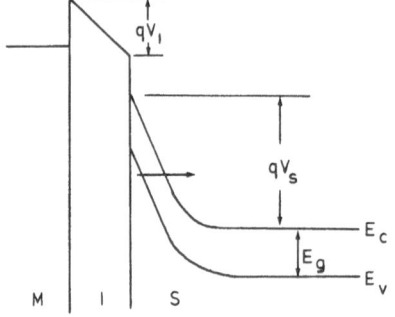

Fig. 3.9. Energy band diagram for a *p*-channel MIS device in deep depletion

Band-to-band tunneling currents can arise in MIS structures on narrow band gap intrinsic materials as a result of the band bending at the semiconductor surface when a potential well is formed. An energy band diagram for an MIS device on an *n*-type substrate is shown in Fig. 3.9. Note that for the amount of band bending shown, tunneling becomes possible between the valence band and the conduction band. Since the minimum width of the barrier to tunneling is $E_g/\mathscr{E}_s$, where E_g is the energy gap and $\mathscr{E}_s$ is the field in the semiconductor, tunneling can be significant for the combination of narrow band gap and high fields. A model for tunneling in narrow band gap materials (such as InSb and HgCdTe) has been formulated by *Anderson* [3.19], based on the $k \cdot p$ theory [3.20]. The result of Anderson's calculation is that the tunneling current density J_t is given by

$$J_t/q = \frac{(qV_s - E_g/2)qF_s}{2^{3/2}\pi^3 h^2} \times \frac{3^{1/2}h}{2P} \times \frac{2(b/a)^{1/2}\exp(-C)}{4+C}, \tag{3.21}$$

where V_s is the magnitude of the surface potential ϕ_s, E_g is the energy gap, F_s is the magnitude of the field in the semiconductor $\mathscr{E}_s$, P is the interband matrix element of the $k \cdot p$ theory, and where

$$b/a = (qV_s - E_g/2)/qV_s, \tag{3.22}$$

$$C = \frac{(a/b)^{1/2}\pi}{2^{3/2}qF_s\hbar} \times \frac{3^{1/2}\hbar}{2P} \times E_g^2. \tag{3.23}$$

The field in the semiconductor varies with the surface potential and so is a function of time for an MIS CCD or CID structure. The maximum field occurs at the surface when the potential well is empty, and can be written for a p-channel device as

$$\mathscr{E}_s(\text{max}) = -Q_D(0)/\varepsilon_s\varepsilon_0 = qN_Dw(0)/\varepsilon_s\varepsilon_0, \tag{3.24}$$

where $Q_D(0)$ is the charge in the depletion region with an empty well, N_D is the substrate donor density, $w(0)$ is the depletion width for an empty well, ε_s is the semiconductor dielectric constant, and ε_0 is the permittivity of free space. Using the depletion approximation, the maximum field can be expressed in terms of the surface potential ϕ_s by

$$\mathscr{E}_s(\text{max}) = -(-2qN_D\phi_s/\varepsilon_s\varepsilon_0)^{1/2}. \tag{3.25}$$

The maximum field value will be used in the tunneling calculation of (3.21) in order to estimate the worst case situation.

The value of the interband matrix element P has been measured as 1.44×10^{-28} joule-m for InSb [3.21]. For $Hg_{1-x}Cd_xTe$ values from 1.25 to 1.39×10^{-28} joule-m, with the most common value being 1.34×10^{-28} joule-m, have been reported [3.22–25]. Since the most common value is also approximately the average of reported results, it will be used here for numerical calculation.

The magnitude of tunneling currents depends so strongly on the band gap, the field, and the matrix element that calculations often relate poorly to experimental results. For the particular case of the narrow gap semiconductors like InSb and HgCdTe, however, the parameters appear to be reasonably well known and *Anderson* has carefully considered the mathematical approximations involved in obtaining (3.21). In addition, the results of the tunneling calculation appear to agree with preliminary measurements on HgCdTe MIS capacitors [3.26].

The tunneling current for an InSb MIS device has been calculated assuming $E_g = 0.24$ eV, and $\varepsilon_s = 17$. The current is plotted in Fig. 3.10 vs substrate doping with the surface potential as a parameter. This curve illustrates the strong dependence of the tunnel current on substrate doping. In InSb MIS devices, dark currents in the range of 100 nA/cm^2 have been measured due to generation [3.18]. According to Fig. 3.10, for a substrate doping of $1 \times 10^{15} \text{ cm}^{-3}$, a surface potential of about 3 V in magnitude would result in a

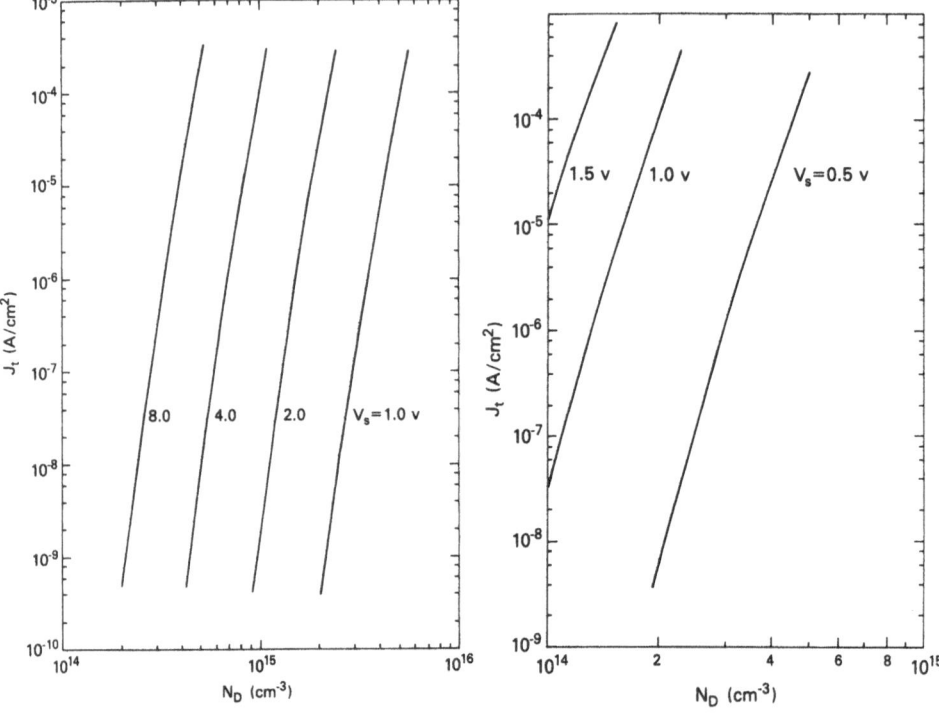

Fig. 3.10. Tunneling current density vs substrate doping with surface potential as a parameter for InSb

Fig. 3.11. Tunneling current density vs substrate doping with surface potential as a parameter for 0.10 eV $Hg_{1-x}Cd_xTe$

tunnel current larger than the generation current. For substrate doping in the mid -10^{15} range, the tunneling currents should be totally dominant for useful surface potentials. For narrower band gaps, the tunneling current becomes much larger. In the case of HgCdTe with $E_g = 0.1$ eV, the calculated current is shown in Fig. 3.11. Even considering the high background which arrays with this band gap would normally see, the tunneling current can clearly be a major problem for MIS devices in 0.1 eV HgCdTe.

Aside from variations in the band gap, reduction in the operating temperature of an MIS device will not result in any reduction in the dark current due to tunneling. It should also be noted that tunneling into surface states or into defect states may also occur in addition to the band-to-band tunneling discussed above.

3.3.5 Transfer Efficiency

At low clocking frequencies, the transfer efficiency of a surface channel CCD tends to be dominated by interface state trapping and reemission. The details of the trapping process are complex, and approximations are required to obtain analytical expressions. The analysis here assumes a uniform trap density in energy across the band gap and a constant capture cross section [3.27]. Under those conditions, the number of carriers lost from a charge packet being transferred into an empty well is

$$N_L = (kT/q)N_{fs}\ln(pn_z+1),\tag{3.26}$$

where N_{fs} is the density of interface states per eV, p is the number of CCD phases per bit, and n_z is the number of empty charge packets transferred through the well prior to the packet containing charge.

In the limiting case of a single empty packet preceding a completely full packet, the transfer efficiency can be written as

$$\varepsilon = qN_L/C_0V_G.\tag{3.27}$$

Combining (3.26) and (3.27) gives

$$\varepsilon = kTN_{fs}\ln(p+1)/C_0V_G.\tag{3.28}$$

A CCD in an FPA may be a two-phase device in order to minimize the number of clock lines on the chip, so $p=2$ may be typical. For a dielectric constant of around 4, an insulator thickness of 1000 Å, and an effective gate voltage of 2 V, the following values can be generated for 77 K:

N_{fs} [cm^{-2} eV^{-1}]	ε
5×10^{11}	8×10^{-3}
1×10^{11}	2×10^{-3}
5×10^{10}	8×10^{-4}
1×10^{10}	2×10^{-4}

Currently, the values of N_{fs}, the interface state density, for MIS devices in InSb and HgCdTe are in the 5×10^{10} to 1×10^{11} range. In order to maintain a high MTF, a CCD register must have a value of $N\varepsilon \sim 0.1$ or less, where N is the total number of transfers [3.1]. For a 32-detector CCD register with two phases, $N=64$. As a result $\varepsilon \sim 10^{-3}$ or less would provide useful MTFs. Longer registers would clearly require better transfer efficiency. The charge trapping that results in the transfer inefficiency also causes a noise component that may be unacceptable. The carrying of a bias charge (or fat zero) in the CCD in order to keep the interface states filled at all times can substantially improve the transfer efficiency. This compensation for the interface states is not perfect due to the fact that the potential well is not a perfectly square well [3.27]. However, a bias charge of a few percent of the well capacity can result in a factor of 10 improvement in the transfer efficiency shown in the table above. This bias charge may be provided by the ir background charge or it can be electrically introduced, depending on the configuration of the array. Electrical introduction of a bias charge does result in an additional noise component which may have to be considered.

An upper frequency limit is set on the clock frequency of the CCD by the time required for the transfer of charge from one CCD gate to the next. The charge packet is transported by combination of self-induced drift, diffusion, and fringing-field drift which results from the potential difference on the two gates involved. Initially, the self-induced drift dominates the transfer. After the initial interval, transfer proceeds by diffusion and fringing-field drift. The last quantities of charge (which determine the transfer efficiency) are transferred very slowly by diffusion, and in order to get reasonable transfer times the fringing field is very important. A rigorous analysis of the transfer process requires detailed calculations with numerical techniques [3.28]. Some closed form analytical approximations have been made [3.29–31]. The fringing field available depends entirely on the geometry of the device and its clocking waveforms. For the types of CCDs useful for FPAs, an estimate can be made of 100–200 V/cm. Based on this estimate, one model [3.30] shows that 20 to 30 transit times would be required to ensure that 99.99 % ($\varepsilon \sim 10^{-4}$) of the charge had transferred, where the transit time is given by

$$t_{tr} = L^2/\mu \tag{3.29}$$

with L the diffusion length and μ the surface mobility. If $L = 25\,\mu m$ and $\mu = 300\,cm^2/V\,s$, then $t_{tr} = 2 \times 10^{-8}\,s$, and the time which must be allowed for nearly complete transfer is $\sim 5 \times 10^{-7}\,s$. This would imply a maximum clock rate of roughly 2 MHz without a degradation in the transfer efficiency. Estimates of this type suggest that the CCD transfer efficiency should be adequate for the TDI register speeds required in an FPA but that the higher speed multiplex register may be harder to do.

3.3.6 Noise Analysis

With a few exceptions (such as TE cooled requirements) in the systems to which FPAs are likely to be applied, a usual design objective is for the system noise to be dominated by the noise associated with the background photon flux (BLIP operation). Whether or not a particular FPA achieves BLIP operation or not depends both on the array characteristics and on the details of the system operation (scan method, integration time, optics, etc.). In this section we shall discuss the major noise contributions due to the arrays and their closely associated signal processing.

The background photon flux, which in the ir is generally much greater than the signal of interest, has fluctuations (noise) associated with it. These fluctuations can be modeled by assuming that the photons arrive at the array randomly, in which case the noise is a shot noise given by the square root of the background. If each photon produces one charge carrier in the semiconductor, then the shot noise on the charge packet will be given by

$$\bar{n}_B = (N_B)^{1/2}, \tag{3.30}$$

where N_B is the number of carriers generated by the background. The background flux depends on the spectral response of the system and on the f number of the optics as well as on the nature of the scene to be viewed. For a 300 K blackbody with $f/2$ optics, the background flux at 4 μm is about 10^{14} photons/cm^2 s arriving at the focal plane. If the quantum efficiency of the FPA is 60 %, then the effective background flux is about 6×10^{13} photons/cm^2 s. The area of the detector sites depends very much on the system application, and may vary from about 25 μm^2 for certain FLIR applications to several 25 μm^2 for search sets. The integration time for the detector sites also depends on the system details and may vary from as low as 10 μs for some scanned systems to 1 ms or more for some staring systems. For the case of a 25×25 μm^2 detector site and a 20 μs integration time (both chosen to be restrictive but reasonably realistic) the background would be about 8×10^3 carriers per site per integration time, resulting in an equivalent input noise due to the background of about 90 carriers. If the detector sites were 50×50 μm^2 with an integration time of 200 μs, the background would be approximately 3×10^5 carriers with the resulting background noise of about 560 carriers. The equivalent noise voltage depends on the device responsivity. For a CCD, the output node capacitance can be as low as a few tenths of a picofarad. Taking 0.3 pF, the 90 carriers would correspond to roughly 50 μV rms and the 560 carriers would correspond to roughly 300 μV rms. A CID with a reasonable number of elements per column (say 16 or so) will have an output node capacitance in the range of a few picofarads, say 5 pF for example. In this device, the 90 carriers would correspond to roughly 3 μV rms noise and the 560 carriers to roughly 20 μV rms. Obviously, for BLIP operation the noise of all of the amplifiers and other signal processing (over the required bandwidth) must be less than the background noise contribution. The example above illustrates that this can be a more difficult problem for a CID than for a CCD.

In addtition to the background shot noise, both CID and CCD FPAs will have a shot noise on the dark current which is given by the square root of the number of equivalent dark current carriers. In the case of a total dark current density of 100 nA/cm^2, a 25×25 μm^2 site with a 20 μs integration time would have roughly 10 carriers of noise while a 50×50 μm^2 site with a 200 μs integration time would have roughly 60 carriers of noise due to the dark current.

All of the FPA readout techniques (and some input techniques when separate detector sites are used) involve reset operations in which a switch is closed to reset a capacitance to some reference voltage. Due to the resistance of the switch, thermal noise is present which results in an uncertainty (noise) associated with the resulting voltage across the capacitor. In terms of rms noise carriers, this uncertainty is given by [3.32]

$$\bar{n}_R = (kTC)^{1/2}/q, \tag{3.31}$$

where C is the capacitance of the node being reset. In many cases, it is possible to eliminate the reset noise by the use of correlated double sampling [3.33]. In some cases, such as the input of an electrical fat zero into a CCD, correlated double sampling cannot be applied and reset noise can be of great concern. Some of the possible CID readout techniques are not compatible with correlated double sampling [3.3] and so may not be useful for FPA applications. As an example, a CID with an output node capacitance of 5 pF operating at 77 K would have a reset noise component of about 450 carriers rms, which could be large enough to dominate the array noise for some applications.

The CID focal plane shown in Fig. 3.2 uses selection switches to connect the appropriate array columns to the readout line. These switches, due to their resistance, have at least a thermal noise which must be considered. The thermal noise voltage is given by

$$\bar{V}_{sw} = (4kTRB)^{1/2}, \tag{3.32}$$

where R is the resistance of the switch and B is the system bandwidth. In the CID, this noise voltage divides between the column capacitance and the total readout line capacitance so that its effect on the output may be reduced by a factor of 2 or so. This noise can be minimized by using low resistance switches (large W/L) and using minimum necessary bandwidths. Some CID focal planes do not use selection switches at all, and avoid this noise source.

In a surface-channel CCD, the fast interface states trap and emit carriers in a random fashion. This process results in an uncertainty (noise) in the charge packet at any time. The statistics of this process result in a carrier noise of [3.32]

$$\bar{n}_{fs} = (pAkTN_{fs}\ln 2/q)^{1/2}, \tag{3.33}$$

where p is the number of phases, A is the CCD gate area, and N_{fs} is the fast interface state density. For the case of a two-phase CCD with a $50 \times 12\,\mu m^2$ gate, operating at 77 K, the values below can be generated from (3.33).

N_{fs} [states/cm^2 eV]	$\bar{n}_{fs}$ [rms carriers]
5×10^{11}	170
1×10^{11}	80
5×10^{11}	55
1×10^{10}	25

Great care must be used in comparing the interface state noise to the background noise to ensure that the proper background is chosen based on system requirements for resolution, MTF, etc.

The array itself is followed by a preamplifier in almost every case. If this preamplifier has a sufficient gain, it will dominate the noise resulting from the rest of the signal processing. Similarly, the noise of the preamplifier will tend to be dominated by the noise of the first stage. There are numerous possible amplifier configurations and designs, of course, but the use of MOS FET devices is attractive from the standpoint of process compatibility with CCD and CID structures. The major components of MOS FET noise are $1/f$ noise and thermal (Johnson-Nyquist) noise in the channel. The $1/f$ noise is proportional to the fast interface state density N_{fs} and inversely proportional to the gate area. Empirically, it is also inversely proportional to some power of the oxide thickness. It is necessary to consider the particular system in order to determine the desired $1/f$ noise corner.

Assuming that the $1/f$ noise corner can be made sufficiently-low, the MOS FET noise due to thermal noise in the channel is given by

$$\bar{V}_c = (8kT/3g_m)^{1/2} \, \text{v/Hz}^{1/2}, \tag{3.34}$$

where g_m is the transconductance of the transistor and is given by

$$g_m = [(2\mu\varepsilon_i\varepsilon_0 W/tL)I_D]^{1/2} \, \text{mhos}, \tag{3.35}$$

where μ is the surface mobility, ε_i is the dielectric constant of the insulator, t is the insulator thickness, W is the gate width, L is the gate length, and I_D is the dc drain current.

Combining (3.34) and (3.35),

$$\bar{V}_c = (8kT/3)^{1/2}(2\mu\varepsilon_i\varepsilon_0/t)^{-1/4}(W/L)^{-1/4}I_D^{-1/4}. \tag{3.36}$$

This equation illustrates in part the trade-off that may be required between transistor bias current (and so power dissipation) and transistor geometry in order to obtain acceptable noise performance at acceptable power dissipation and device area. Large W/L ratios also tend to result in large gate capacitance. Since this gate capacitance of the first preamplifier stage is connected to the output node, it can result in a lowered array responsivity. A general rule of thumb is that the transistor geometry can be increased until the gate capacitance is approximately equal to the output node capacitance. In the case of n-type InSb at 77 K with an insulator having a dielectric constant of 4, the following numbers are typical:

$$\mu = 300 \, \text{cm}^2/\text{V} \cdot \text{s}$$

$$t = 1000 \, \text{Å}$$

and (3.36) can be written as

$$\bar{V}_c = 7.8 \times 10^{-10} (W/L)^{-1/4} I_D^{-1/4} . \tag{3.37}$$

For illustration purposes, a bias current of $100\,\mu A$ and a W/L ratio of 50 would result in $\bar{V}_c = 3\,nV/Hz^{1/2}$. In order to reach $2\,nV/Hz^{1/2}$ without changing W/L, the drain current would have to be increased to about $500\,\mu A$ (with corresponding increase in power dissipation). Similarly, to reach $2\,nV/Hz^{1/2}$ without changing the current would require an increase in W/L to about 250 (increasing the device size significantly and increasing the input gate capacitance). The noise performance actually required of the amplifier depends directly on the amplifier bandwidth as defined by the system. In the previous discussion on noise, a value of $3\,\mu V$ of background noise was calculated for a CID under one set of conditions. If the amplifier noise is set to match the background then it too would be $3\,\mu V$. If the system design requires a bandwidth of 1 MHz, then the transistor noise must be limited to $3\,nV/Hz^{1/2}$ (which is a very low noise device). On the other hand, if the system design requires a bandwidth of only 100 kHz, then the transistor noise must only be limited to $9\,nV/Hz^{1/2}$ (which is reasonably straightforward to do).

3.3.7 Interface Circuits

Intrinsic FPAs of the type shown in Fig. 3.1b perform the detection in the CCD register itself, and so have no interface circuit required between the detector and the CCD TDI or readout register. Arrays of the type in Fig. 3.1c, however, do require an interface between the detector and the CCD. That interface may be as simple as a transfer gate to control the flow of charge from an MIS detector into the CCD or as elaborate as coupling circuit for a PV detector with background suppression and a buffer amplifier included. The complexity required in the interface circuit depends largely on the characteristics of the detector, the size of the background signal, and the dynamic range which can be accommodated by the CCD.

As an illustration of the considerations required, some of the aspects of interfacing a PV detector to a CCD will be discussed. It must be pointed out that advanced FPAs involve hundreds to thousands of detectors per chip and several such chips per focal plane. In order to achieve acceptable power dissipation and packing density, the power consumption and size of any interface circuits must be strictly limited. Two basic techniques are available (each having numerous variations) for interfacing PV detectors to CCDs. Figure 3.12a illustrates the source modulation or direct injection techniques, and Fig. 3.12b illustrates the gate modulation technique.

Direct injection involves the modulation of the source of the CCD input FET (formed by the input diode as a source, G_1 as a gate, and the potential well under G_2 as a drain) by the signal from the PV detector. The virtue of direct

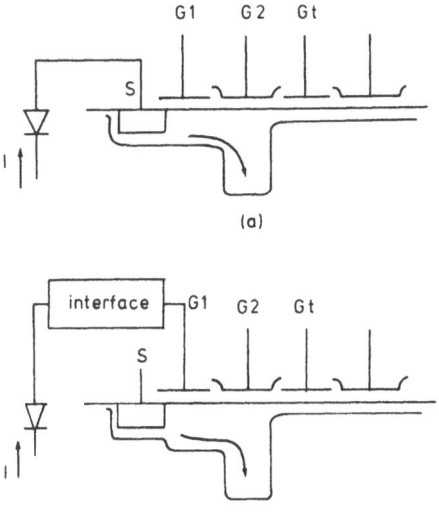

(a)

(b)

Fig. 3.12a, b. Basic interface techniques for hybrid arrays: (a) source modulation and (b) gate modulation

injection is its relative simplicity – with the low power and small size which result. The major problems associated with this interface technique are 1) uniformity of the detector operating points across an array, 2) the injection efficiency of the input, 3) the frequency response of the input circuit (for scanned arrays), and 4) the signal handling capacity required in the CCD. The operating point of the PV detector is determined by the characteristics of the input FET. At the typical current levels involved for ir PV detectors, the input FET operates in the subthreshold region. Variations in the effective threshold voltage of the input FETs across an array result in variations in the bias points of the detectors. Typical processing of MIS devices in silicon results in threshold variations in the range of 30 to 100 mV. Processing of MIS devices on ir semiconductors may very well result in larger variations. These bias variations can result in the input FET being completely cut off, in the one extreme, or in the detector being operated at a relatively large reverse bias (resulting in excess dc current, poor injection efficiency, and perhaps excess $1/f$ noise), in the other extreme. Very careful processing of the MIS devices, and the fabrication of diodes tolerant of large reverse bias are required to reduce this problem for direct injection inputs. The signal current from the PV detector is divided between the diode incremental resistance and the incremental input resistance of the input FET. The ratio of the current into the FET to the total current is known as the injection efficiency. A high injection efficiency leads to the minimum degradation of the signal-to-noise ratio by the input circuit. High injection efficiency results from diodes with very large resistance and from input FETs with large values of transconductance g_m (since the input resistance of the FET is approximately $1/g_m$). Obtaining a large value of g_m is difficult since the input FET usually operates subthreshold and since large W/L ratios are not compatible with size constraints. High-resistance diodes require very careful control of all surface and bulk generation or leakage currents. The exact values required for the diode resistance and FET transconductance depend critically on the specific system details, but for typical g_m values obtainable, the diode resistance-area product needed for 3–5 µm backgrounds is in the range of several thousand ohm-cm^2 and for 8–12 µm backgrounds is in the range of tens of ohm-cm^2. In scanned arrays, the motion of the scene across the array results in signal frequencies up to 30 kHz or so, depending on the system. The input circuit must have a frequency response suitable for these applications. A high-frequency response implies a short input time constant, which in turn means low detector capacitance and low FET input resistance (high g_m). This frequency response can be difficult to achieve as a result of g_m limitations. The direct injection technique passes basically the full background signal from the detector into the CCD. As a result, CCD charge capacity problems can arise, especially for TDI processing. These saturation problems must be carefully considered for 3–5 µm arrays at tactical backgrounds and almost certainly rule out direct injection alone as an input technique for 8–12 µm arrays. Some methods have been proposed for adding background suppression to direct injection inputs. One such method is charge splitting, in which the charge packet collected by the drain of the input FET is physically divided, with some fixed fraction of the packet going into the CCD while the rest is dumped to a charge sink. A second method involves continuously sinking a fraction of the current, with the fraction selected adaptively on the basis of the average background from the scene. More details concerning direct injection and techniques for g_m enhancement (including the use of a buffer amplifier) can be found in [3.34–36].

Gate modulation involves coupling the signal from the detector to the gate of the input FET. The primary interest in this technique is its use in various types of background suppression inputs. It does not suffer from injection efficiency problems, but it does have the same frequency response constraints as direct injection. In addition, the gate modulation technique requires the use of some form of load resistor for the detector. One possible background suppression technique for gate modulation is the simple ac coupling circuit shown in Fig. 3.13 [3.37]. A small capacitor provides ac coupling and an FET switch is used to periodically reset the input node. Gate modulation is very sensitive to threshold variations among the input FET gates. Some type of potential equilibration input can be used to reduce this effect. Further discussion of gate modulation schemes is available in [3.34, 38].

3.4 FPA Developments

Several development programs in intrinsic FPAs are underway. One approach using an InSb CID technology has resulted in the fabrication of both linear and area arrays [3.39, 40]. These devices are fabricated using an SiON deposited

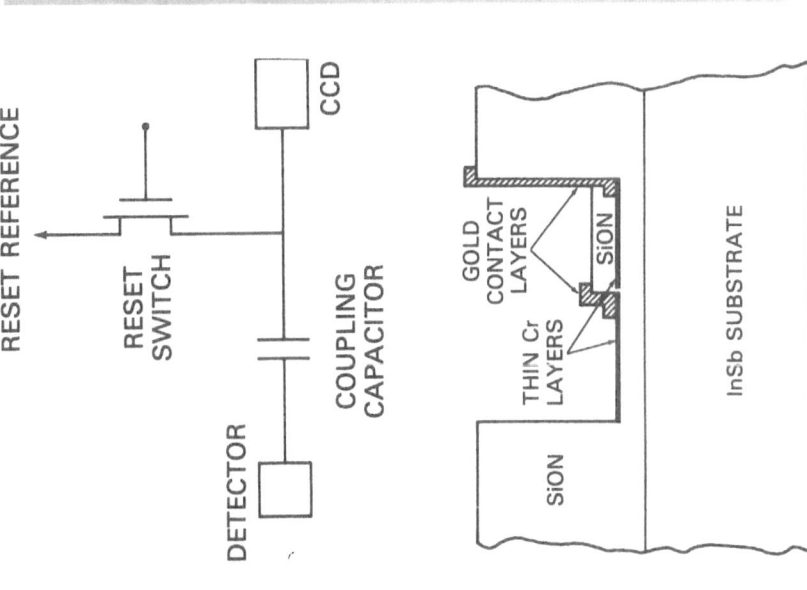

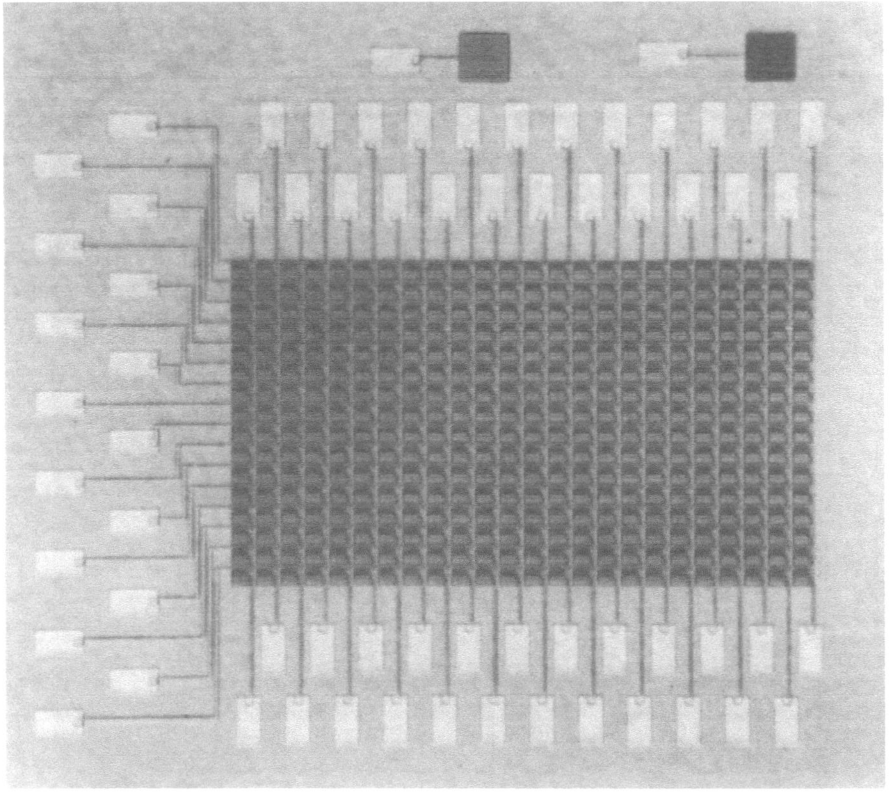

Fig. 3.15. Photograph of a 16 × 24 element InSb CID array

RESET REFERENCE

RESET
SWITCH

CCD

DETECTOR

COUPLING
CAPACITOR

GOLD
CONTACT
LAYERS

SiON

THIN Cr
LAYERS

InSb SUBSTRATE

SiON

Fig. 3.14. Cross section of one InSb CID array site showing fabrication

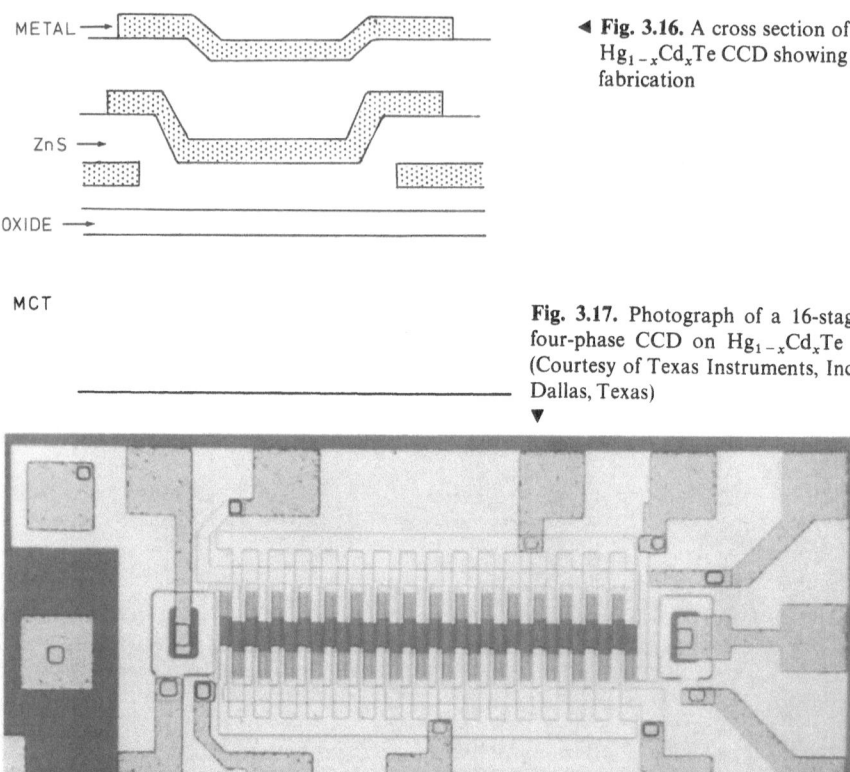

METAL →

ZnS →

OXIDE →

◀ **Fig. 3.16.** A cross section of a $Hg_{1-x}Cd_xTe$ CCD showing fabrication

MCT

Fig. 3.17. Photograph of a 16-stage four-phase CCD on $Hg_{1-x}Cd_xTe$ (Courtesy of Texas Instruments, Inc., Dallas, Texas)
▼

insulator, with thin chrome as the transparent gate material and thick gold for all other gates and interconnects. A cross section of one such detector site is shown in Fig. 3.14. A 16×24 element array fabricated by this process is shown in Fig. 3.15. The detector sites in this array are $50 \times 50 \,\mu m^2$ with a $66 \,\mu m$ pitch in the 16-element direction and a $75 \,\mu m$ pitch in the 24-element direction. This array was designed for use in a scanned focal plane in conjunction with a silicon signal processor chip which provides preamplifiers, TDI, ac coupling, and multiplexing functions. Details of the fabrication and operation of these arrays have been reported [3.18, 40]. Good arrays have dark current densities in the range of $100 \,nA/cm^2$ at 77 K, and staring mode measurements have shown total noise of about 200 carriers for 1 ms integration times at $4.2 \,\mu m$ wavelength. A 1×32 array has been used for scanned imaging. A program is under way to apply 16×64 element arrays of this type to the assembly of a 525-line, TV compatible, vertically scanned FLIR having about 8000 detectors.

An InSb CCD development program has resulted in transfer efficiencies of at least 0.99 for 20-stage devices, as well as the successful operation of InSb MIS

FET devices [3.41]. In addition, the fabrication of CCD arrays on such materials as InAsSb and InGaSb in order to take advantage of the variable cutoff wavelength has been pursued [3.42].

A CCD in 3–5 µm HgCdTe has been fabricated [3.43] which has shown transfer efficiency greater than 0.999. These devices have been fabricated with thin metal gates and have been operated to demonstrate TDI for a 16-stage register. Figure 3.16 shows a typical cross section and Fig. 3.17 shows a completed device.

In addition to the monolithic arrays mentioned, a number of intrinsic hybrid arrays have been fabricated with PV detectors interfaced to silicon CCD processors, typically by metal bump techniques [3.34] or by evaporated lead techniques [3.44]. Materials used so far include InSb, PbSnTe, InGaSb, InAsSb, and HgCdTe. Arrays up to 32×32 elements have been fabricated with good diode yields and good interconnect yields.

There seems to be little doubt that arrays of several thousands of detectors will become available for a wide variety of applications within the next few years. The ultimate configurations and materials will depend on the rate of advance in the various approaches and on the details of the eventual applications.

References

3.1   D.F.Barbe: Proc. IEEE **63**, 38 (1975)
3.2   D.F.Barbe: IEEE J. SSC-**11**, 109 (1976)
3.3   H.F.Burke, G.J.Michon: IEEE J. SSC-**11**, 121 (1976)
3.4a A.F.Milton, M.Hess: "Series-Parallel Scan IR CID Focal Plane Concept", in Proc. 1975 Intern. Conf. Application of Charge-Coupled Devices, p. 71
3.4b R.J.Keyes (ed.): *Optical and Infrared Detectors*, Topics in Applied Physics, Vol. 19 (Springer, Berlin, Heidelberg, New York 1977) p. 220
3.5   R.D.Thom, R.E.Eck, J.D.Phillips, J.B.Scorso: InsSb CCDs and Other MIS Devices for Infrared Applications", in Proc. 1975 Intern. Conf. Application Charge-Coupled Devices, p. 31
3.6   J.C.Kim: "InSb MIS Technology and CID Devices", in Proc. 1975 Intern. Conf. Application of Charge-Coupled Devices, p. 1
3.7   D.Seib: IEEE Trans. ED-**21**, 210 (1974)
3.8   R.K.Willardson, A.C.Beer: "Infrared Detectors", in *Semiconductors and Semimetals*, Vol. 5 (Academic Press, New York) p. 27
3.9   A.S.Grove: *Physics and Technology of Semiconductor Devices* (John Wiley and Sons, New York 1967) p. 271
3.10 S.M.Sze: *Physics of Semiconductor Devices* (Wiley-Interscience, New York 1969) p. 425
3.11 A.F.Tasch, R.A.Chapman, B.H.Breazeale: J. Appl. Phys. **41**, 4202 (1970)
3.12 R.A.Chapman, M.A.Kinch, A.Simmons, S.R.Borrello. H.B.Morris, J.S.Wrobel, D.D.Buss: Appl. Phys. Lett. **32**, 434 (1978)
3.13 E.H.Putley: *The Hall Effect and Related Phenomena* (Butterworth, London 1960) p. 113
3.14 J.L.Schmit: J. Appl. Phys. **41**, 2876 (1970)
3.15 J.L.Schmit, E.L.Stelzer: J. Appl. Phys. **40**, 4865 (1969)
3.16 M.A.Kinch, S.R.Borrello: Infrared Phys. **15**, 111 (1975)
3.17 D.K.Schroder, J.Guldberg: Solid-State Electron. **14**, 1285 (1971)

3.18 J.C.Kim, W.E.Davern, D.Colangelo: "Continued Development of Indium Antimonide CID Arrays", Final Technical Report on Contract N00173-76-C-0128, General Electric, Syracuse (1977)

3.19 W.W.Anderson: Infrared Phys. **17**, 147 (1977)

3.20 E.O.Kane: "The $k \cdot p$ Method", in *Semiconductors and Semimetals*, Vol. 1, ed. by R.K.Willardson, A.C.Beer (Academic Press, New York 1966) p. 75

3.21 O.Madelung: *Physics of III–V Compounds* (John Wiley and Sons, New York 1964)

3.22 M.A.Kinch, D.D.Buss: "Far Infrared Cyclotron Resonance in HgCdTe", in *Physics of Semimetals and Narrow Gap Semiconductors*, ed. by D.Carter, R.Bate (Pergamon Press, Oxford 1971) p. 461

3.23 B.D.McCombe, R.J.Wagner, G.A.Prinz: Phys. Rev. Lett. **25**, 87 (1970)

3.24 S.H.Groves, T.C.Harman, C.R.Pidgeon: Solid State Commun. **9**, 451 (1971)

3.25 G.A.Antcliffe, R.T.Bate, R.A.Reynolds: "Oscillatory Magnetoresistance from an *n*-type Inversion Layer with Nonparabolic Bands", in *Physics of Semimetals and Narrow Gap Semiconductors*, ed. by D.Carter, R.Bate (Pergamon Press, Oxford 1971) p. 499

3.26 D.R.Rhiger, J.D.Langan: "Study of HgCdTe MIS Technology", Final Technical Report on Contract N00173-76-C-0316, Santa Barbara Research Center, Santa Barbara (1977)

3.27 C.Sequin, M.Tompsett: *Charge Transfer Devices* (Academic Press, New York 1975)

3.28 R.J.Strain, N.L.Schryer: Bell System Tech. J. **50**, 1721 (1971)

3.29 D.B.Scott, S.G.Chamberlain: IEEE Trans. SSC-**12**, 45 (1977)

3.30 H.S.Lee, L.G.Heller: IEEE Trans. ED-**19**, 1270 (1972)

3.31 J.E.Carnes, W.F.Kosonocky, E.G.Ramberg: IEEE Trans. SSC-**6**, 323 (1971)

3.32 J.E.Carnes, W.F.Kosonocky: RCA Rev. **33**, 327 (1972)

3.33 M.H.White, D.R.Lampe. F.C.Blaha, I.A.Mack: IEEE Trans. SSC-**9**, 1 (1974)

3.34 J.T.Longo, D.T.Cheung, A.M.Andrews, C.C.Wang, J.M.Tracy: IEEE Trans. ED-**25**, 213 (1978)

3.35 A.J.Steckl: "Injection Efficiency in Hybrid IR CCDs", in Proc. 1975 Intern. Conf. Application Charge-Coupled Devices, p. 85

3.36 N.Bluzer, R.Stehlik: IEEE Trans. ED-**25**, 160 (1978)

3.37 S.P.Emmons, T.F.Cheek, J.T.Hall, P.W.Van Atta, R.Balcerak: "A CCD Multiplexer with Forty AC Coupled Inputs", in Proc. 1975 Intern. Conf. Application Charge-Coupled Devices, p. 43

3.38 W.Grant, R.Balcerak, P.Van Atta, J.T.Hall: "Integrated CCD-Bipolar Structure for Focal Plane Processing of IR Signals" in Proc. 1975 Intern. Conf. Application Charge-Coupled Devices, p. 53

3.39 J.C.Kim, W.E.Davern, D.Colangelo: Proc. 1976 IEEE Intern. Electron Devices Meeting, p. 550

3.40 J.C.Kim: IEEE Trans. ED-**25**, 232 (1978)

3.41 R.D.Thom, F.J.Renda, W.J.Parrish, T.L.Koch: Proc. 1978 IEEE Intern. Electron Devices Meeting, p. 501

3.42 E.E.Barrowcliff, L.O.Bubulac, D.T.Cheung, A.M.Andrews, J.D.Blackwell, F.Fox, E.R.Gertner, W.E.Tennant, M.J.Ludowise, L.E.Wood: "Planar GaInSb CCDs", in Proc. 1978 Intern. Conf. Application of Charge-Coupled Devices, pp. 2–77

3.43 D.D.Buss, R.A.Chapman, M.A.Kinch, S.R.Borrello, A.Simmons, C.G.Roberts: Proc. 1978 IEEE Intern. Electron Devices Meeting, p. 496

3.44 R.Broudy, M.Rheine: "Advances in Hg CdTe Infrared Focal Plane Technology", SPIE Proc., Vol. 124 (1977) p. 62

4. Extrinsic Silicon Focal Plane Arrays

D. K. Schroder

With 27 Figures

Although extrinsic Si infrared detectors have been known for 28 years, it is only within the last 5 years that significant progress has been made in the development of extrinsic monolithic focal plane arrays (MFPAs). The main driving force behind this technology is the fact that both detection and signal processing functions can be implemented on a common Si chip.

Device and materials properties are treated in this chapter. Considerations that are of secondary importance in single detectors become very important in MFPAs, for example free carrier absorption, optical and electrical cross talk. Neutron transmutation doping is described as a method to achieve very precise compensation.

Problems specific to MFPAs are the injection structures and several possibilities are discussed. The layouts considered in today's devices are outlined and finally the requirements for higher temperature operation are given.

4.1 Historical Background

The photoconductive effect was discovered by *Smith* [4.1] in 1873 when he experimented with selenium as an insulator for submarine cables. It was not until 1917, however, that Case demonstrated a thallous sulfide detector using this effect. Serious efforts to develop infrared (ir) detectors began just prior to World War II in Germany and the United States [4.2]. Many materials have been investigated since the early "thallous sulfide days" and it is interesting to look at the materials development with time, because it alternates between intrinsic and extrinsic, as shown in Fig 4.1. The dates shown there are approximate starting dates of significant development efforts, leading in all cases to usable detectors. In fact, most of the materials shown there are still in use today. For example, lead sulfide, one of the earliest detector materials on that list, is used in more applications today than at any time in its history. Furthermore, for most of the materials shown, improvements are still being made today in both materials growth and detector fabrication. However, the major emphasis today is shifting from the perfection of new materials to techniques of information processing on the focal plane. This is because materials are approaching their theoretical limits, while sophisticated signal processing is only beginning to be exploited.

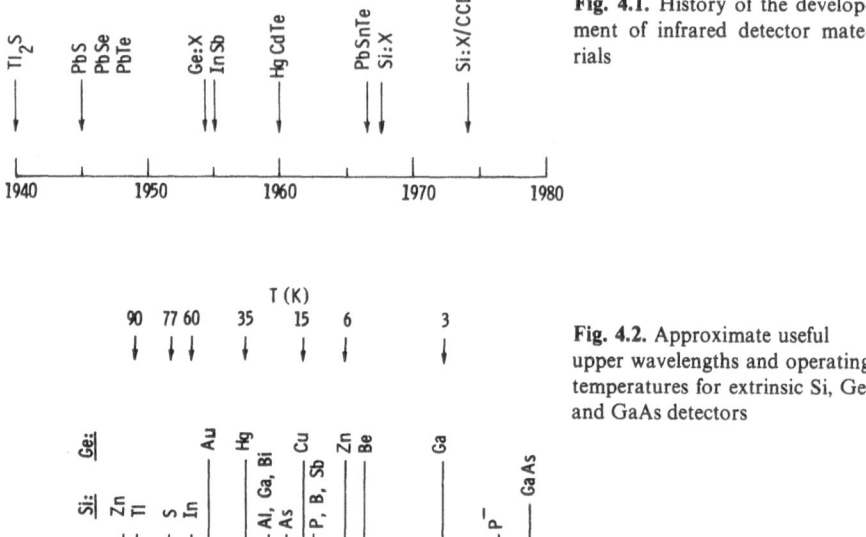

Fig. 4.1. History of the development of infrared detector materials

Fig. 4.2. Approximate useful upper wavelengths and operating temperatures for extrinsic Si, Ge, and GaAs detectors

Of the extrinsic materials, only germanium (Ge) and silicon (Si) have been developed to any appreciable extent, although some work was done on gallium arsenide [4.3]. Work on Si was first reported in 1951 by *Burstein* et al. [4.4] in the United States, to be followed by a publication from England in 1952 [4.5]. A year later the first results on Ge were reported [4.6]. Because Ge was easier to purify than Si, higher performance detectors could be achieved and the emphasis in extrinsic detectors shifted to Ge. Various impurities were studied and the wavelength range to 120 μm can be covered with these detectors, as shown in Fig. 4.2.

During the early 1960s extrinsic Si with various impurities was studied in the USSR [4.7] and in 1967 the first comprehensive detector-oriented paper was published by *Soref* [4.8]. However, the state of extrinsic Si was not changed significantly. Although Si has several advantages over Ge, namely lower dielectric constant giving shorter dielectric relaxation time and lower capacitance, higher dopant solubility and larger photoionization cross section for higher quantum efficiency, and lower refractive index for lower reflectance, these were not sufficient to warrant the necessary development efforts needed to bring it to the level of the, by then, highly developed Ge detectors. However, there was one early application of extrinsic Si, namely, as an ir-sensitive vidicon, using a gold-doped Si target. It had to be cooled to 77 K, but was sensitive to only about 2 μm [4.9].

After being dormant for about 10 years, extrinsic Si was reconsidered after the invention of the charge-coupled device (CCD). Instead of the traditional individual detector with amplifier, it became possible to have much more sophisticated readout schemes. For the first time both detection and readout functions could be implemented on one common chip without elaborate interconnection schemes. This has not been possible with other semiconductor materials because their development has not yet progressed to the point where signal processing functions can be successfully fabricated on the same chip as the detector.

It appears then, at this time, that extrinsic Si detectors will play a significant role in future ir imaging systems, because many detectors can be integrated on a chip and various signal processing functions can be performed on the focal plane in a monolithic focal plane array (MFPA) [4.9a]. This should lead to higher performance at lower cost.

In the first part of this chapter, extrinsic Si materials and detector properties are briefly discussed as they pertain to the performance of MFPAs. For a more detailed discussion of extrinsic Ge and Si the reader is referred to the excellent review article by *Bratt* [4.10]. In the second part, the integration of detection and signal processing functions on a common chip is addressed with emphasis on techniques of injecting the photocharge into the readout circuit. This is followed by some examples and results and some speculations about the future of these devices.

4.2 Detector and Materials Considerations

4.2.1 Photocurrent

In the following, we shall consider a p-type extrinsic detector with N_A acceptors, N_D compensating donors and N_B boron impurities; all units in cm^{-3}. The boron is included since it is always present in Si owing to its high segregation coefficient. The three impurities are shown in Fig. 4.3.

The material is further characterized by the lifetime τ, the mobility μ, and the quantum efficiency η. The thermal (p_{th}) and optical (p_{op}) hole densities, the detector dimension l along the optical path length, and the detector area A are also important.

There are Q_B photons/s cm^2 incident on the detector, as shown in Fig. 4.4. The photocurrent is given by

$$i_{ph} = q p_{op} v A . \tag{4.1}$$

With

$$p_{op} = \eta Q_B \tau / l \tag{4.2}$$

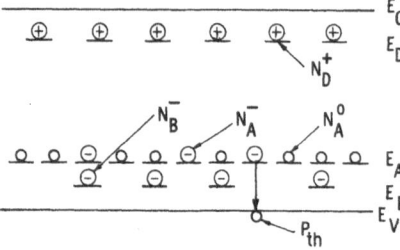

Fig. 4.3. p-type extrinsic photoconductor energy band diagram

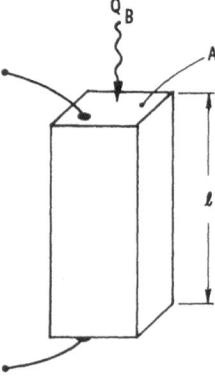

Fig. 4.4. Detector configuration showing the area A and length l. The photons are incident from the top and contacts are at the top and bottom

and

$$v = \mu\mathscr{E} \tag{4.3}$$

the photocurrent can be written as

$$i_{ph} = q\eta G_0 Q_B A , \tag{4.4}$$

where the photoconductive gain G_0 is

$$G_0 = \mu\tau\mathscr{E}/l . \tag{4.5}$$

It is clear from (4.4) that a high photocurrent requires high mobility, long lifetime, and as short a detector as is consistent with a high quantum efficiency, to be discussed later on.

4.2.2 Photoconductive Gain

The photoconductive gain of (4.5) is shown as a constant, i.e., independent of frequency. This is not correct since extrinsic detectors are subject to gain saturation due to carrier sweepout and dielectric relaxation. In the following discussion, we shall follow *Bratt* [4.10].

Consider the detector of Fig. 4.3, subjected to a short light pulse. The pulse will produce p_{op} holes and an equal density of negatively charged acceptors N_A^-. The holes are swept out of the detector in a transit time, leaving behind a uniform distribution of ionized acceptors. It is assumed here that the drift length $l_d = \mu\tau\mathscr{E}$ is larger than the detector length l. The detector relaxes back to its neutral state within the dielectric relaxation time

$$\tau_\varrho = K\varepsilon_0\varrho, \tag{4.6}$$

where K is the dielectric constant and ϱ the resistivity of the detector.

The dielectric relaxation frequency for Si is given by

$$f_\varrho = 1/2\pi\,\tau_\varrho = q\eta\mu\tau Q_B/2\pi\,K\varepsilon_0 l$$
$$= 2.5 \times 10^{-8}\eta\mu\tau Q_B/l\,[\text{Hz}]. \tag{4.7}$$

Using $\eta = 0.3$, $\mu = 8 \times 10^3\,\text{cm}^2/\text{V s}$, $\tau = 10^{-8}\,\text{s}$, and $l = 0.05\,\text{cm}$ gives

$$f_\varrho \simeq 1.2 \times 10^{-11}Q_B\,[\text{Hz}].$$

For low background applications, where $Q_B \simeq 10^{12}\,\text{photons/s cm}^2$, f_ϱ is only 12 Hz, while even for conventional 300 K terrestrial imaging, f_ϱ is only in the low kHz range.

Dielectric relaxation time effects are observed when holes are swept out of the detector without replenishment from the contacts. It is obvious then that the photoconductive gain should be frequency dependent. There are two models that describe this frequency dependence, both shown in Fig. 4.5. The model of [4.11], shown by the dot-dash curve, predicts a gain drop at f_ϱ, while that of [4.12] (solid curve), predicts a corner frequency of $f_\varrho/2Q_0$, where G_0 is the low-frequency gain given by (4.5). The model of [4.11] is verified by experiments on both extrinsic Ge and Si. If there were no dielectric relaxation effects, the gain should follow the dashed curve, where the high-frequency corner is determined by lifetime or circuit time constant limitations.

The gain G_0 of most present extrinsic Si detectors is of the order of unity, because of the low lifetimes achieved thus far. Hence, the arguments above are

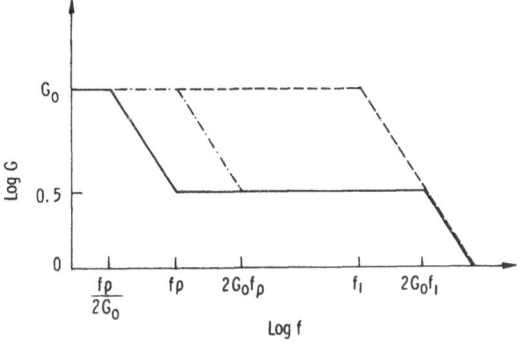

Fig. 4.5. The frequency behavior of the photoconductive gain. The various models are described in the text

not very important and using a frequency-independent gain is reasonable. However, with material improvements, lifetime improvements can be expected with resultant gain increases. Then the frequency dependence of the gain will need to be considered.

4.2.3 Thermal Carrier Concentration

The $D*$ expression for a photoconductor is given by [4.13]

$$D* = \frac{\eta\lambda}{2hc}\sqrt{\frac{\tau}{(p_{th}+p_{op})l}}. \tag{4.8}$$

It is a generally used figure of merit because it contains both the responsivity and noise behavior of the detector. Maximum $D*$ is achieved when $p_{th} \ll p_{op}$, i.e., the thermal hole concentration should be reduced for the detector to be dominated by optically generated carriers.

This is shown graphically in Fig. 4.6, where $D*$ is plotted against temperature. For $T < T_1$, the detector is BLIP (background limited impurity photoconductor), while for $T > T_1$, $D*$ is determined by p_{th}. The thermal hole concentration is obtained from charge neutrality considerations, which, from Fig. 4.3, give

$$p_{th}+N_D^+ = N_B^- + N_A^-, \tag{4.9}$$

where

$$N_D^+ = N_D/[1+2\exp(E_F-E_D)/kT)]$$
$$N_B^- = N_B/[1+4\exp(E_B-E_F)/kT]$$
$$N_A^- = N_A/[1+4\exp(E_A-E_F)/kT]$$
$$p_{th} = N_v\exp(-E_F/kT).$$

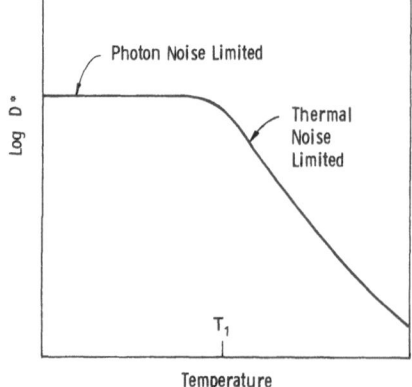

Fig. 4.6. $D*$ as a function of temperature. T_1 is the boundary between photon noise and thermal noise

A solution of (4.9) requires a knowledge of the Fermi level E_F. This can be obtained exactly by iteration techniques and must be used for the most general case. However, for the usual case where all the B atoms are compensated and $p_{th} \ll N_D \ll N_A$, the expression

$$p_{th} = \frac{N_A N_v}{g(N_D - N_B)} \exp(-E_A/kT) \qquad (4.10)$$

gives correct results [4.14]. Here N_v is the density of states in the valence band and g is the degeneracy factor, which is 4 for p-type and 2 for n-type impurities.

The boron impurities are present in Czochralski-grown Si in concentrations of $(3-10) \times 10^{13}$ cm^{-3}, while in float-zone material the density is $(0.1-1) \times 10^{13}$ cm^{-3}. For deep-level dopants like In, with operating temperatures of 50–60 K, the B impurities would each add a thermal hole. Such high p_{th} would make the detector useless and there are two options to reduce p_{th}: (I) reduce the temperature to freeze the holes onto the B, or (II) add compensating donors. The former is clearly undesirable and the latter is used. The effects of such compensation on p_{th} can be seen in Fig. 4.7, plotted for typical In and B concentrations.

It is instructive at this point to digress briefly to consider the reason for the lower operating temperatures of extrinsic detectors, compared to intrinsic photoconductors of equal spectral response. The D^* expression, in the region where thermal carriers dominate, from (4.8) is given by

$$D^* = \frac{\eta\lambda}{2hc} \sqrt{\frac{\tau}{p_{th} l}}. \qquad (4.11)$$

For *extrinsic* detectors, p_{th} is the *majority* carrier density, while for the *intrinsic* case it is the *minority* carrier density. Recall that for intrinsic photoconductors,

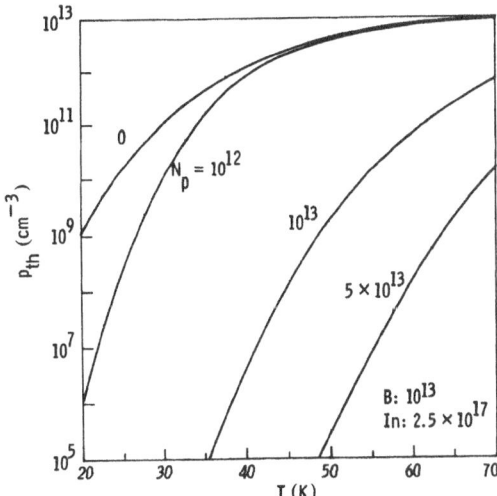

Fig. 4.7. Thermal hole concentration for Si:In with B impurities as a function of compensating donor density. The 0.11 eV level impurity was not considered in these curves

the conduction is controlled by majority carriers but the noise current is primarily due to fluctuations of the minority carriers.

The thermal hole density for extrinsics is given by (4.10). For intrinsics it is

$$p_{th}(\text{intr}) = n_i^2/n_0 = \frac{N_v N_c}{n_0} \exp(-E_G/kT), \qquad (4.12)$$

where N_c is the density of states in the conduction band, n_0 is the majority carrier density, and E_G is the bandgap. For equal wavelength response, $E_A = E_G$. Furthermore, the density of states N_v is similar for the two cases since the hole effective masses are not too different. With these simplifications

$$\frac{p(\text{extr})}{p(\text{intr})} \simeq \frac{N_A n_0}{g(N_D - N_B)N_c}. \qquad (4.13)$$

For $N_A = 10^{17}\,\text{cm}^{-3}$, $(N_D - N_B) = 10^{13}\,\text{cm}^{-3}$, $n_0 = 10^{14}\,\text{cm}^{-3}$, $g = 4$, and $N_c = 3 \times 10^{15}\,\text{cm}^{-3}$ (corresponding to an electron effective mass of approximately 0.01, typical of intrinsic ir materials and $T = 50\,\text{K}$), the ratio in (4.13) becomes roughly 100, showing that the thermal hole concentration of extrinsics is significantly higher than that of intrinsics at a given temperature. In addition, τ is generally lower and l higher in extrinsic detectors. The end result is that the 3 dB temperature T_1 of Fig. 4.6 is typically 30–40 °C lower for extrinsic detectors.

Equation (4.8) indicates that BLIP performance can be achieved, provided $p_{th} \ll p_{op}$. According to (4.10), this can be obtained by a combination of compensation and cooling. There are, however, some complications that can occur. These are impact ionization, hopping, and impurity band conduction, all of which degrade D^* by introducing additional noise components.

Impact ionization is caused by free holes gaining sufficient energy from the applied electric field to ionize neutral impurity atoms. This process not only creates additional free carriers, but the impact ionization process adds noise to the device. It happens at a critical electric field, which increases with increasing majority impurity concentration, because higher concentrations reduce the carrier mobility through neutral impurity scattering. This decreases the rate at which carriers gain energy from the electric field.

When the majority impurity concentration is low, the spacing between atoms is large and there is negligible interaction between impurities. As the concentration is increased and the distance between atoms becomes sufficiently small, carriers can hop from one impurity to another. The probability of hopping is enhanced by compensating impurities, which, by ionizing some of the majority impurities, make empty sites available for carriers to hop into.

For still higher concentrations, the impurity level forms into a band, and conduction takes place by carriers flowing within this band. For both hopping and impurity band conduction, current flows without the need to excite holes into the valence band. Detector performance is degraded by reducing the ratio of photoconductive/dark current and by increasing device noise.

Hopping conduction depends on the energy level of the impurity with its onset occurring at higher concentrations for deeper lying impurities. For Si:Ga, it begins at concentrations in the high $10^{16}\,\text{cm}^{-3}$, while for Si:In, the concentrations are higher by about a factor of 10. By keeping the electric field below the breakdown value and the majority impurity concentrations below their "hopping" or "impurity band" values, all three of these deleterious effects are avoided.

4.2.4 Quantum Efficiency

Consider the detector of Fig. 4.4, with reflectivities R_1 and R_2 at the top and bottom surfaces, respectively. The ratio of the photon flux absorbed within the detector to the incident flux is defined as the quantum efficiency

$$\eta = \frac{(1-R_1)(1-e^{-\alpha l})(1+R_2 e^{-\alpha l})}{(1-R_1 R_2 e^{-2\alpha l})}. \tag{4.14}$$

The photocurrent that flows as a result of the incident photon flux density Q_B is proportional to the product of the quantum efficiency and the probability of carrier collection, and is given by (4.4).

Equation (4.14) reduces to the familiar form

$$\eta = \frac{(1-R)(1-e^{-\alpha l})}{(1-Re^{-\alpha l})} \tag{4.15}$$

for $R_1 = R_2 = R$. The reflectivity can be made small by proper antireflection coatings to give

$$\eta = 1 - e^{-\alpha l}. \tag{4.16}$$

The best case, however, is obtained for $R_1 = 0$ and $R_2 = 1$, i.e., the light-incident surface is antireflection coated for minimum reflection, while the back surface is treated for total reflection. For this case

$$\eta = 1 - e^{-2\alpha l}. \tag{4.17}$$

Although this case is best for highest quantum efficiency, it can introduce optical cross talk by allowing nonabsorbed irradiation to be reflected back into the device, as discussed further on.

The absorption coefficient α is given by

$$\alpha = \sigma_0 N_A^0. \tag{4.18}$$

It is the product of the photoionization cross section σ_0 and the neutral acceptor concentration. It is desirable to make α as large as possible. However, since σ_0 is a property of the impurity which is not under the control of the device designer, the only option left is to increase N_A^0. The upper limit here is set by either "hopping" or "impurity band" conduction, as discussed earlier.

Various attempts have been made to develop theories that can predict the photoionization cross section [4.15]. Some of these are applicable to deep-lying impurities, while others are better suited to impurities with shallow energy levels. The model due to *Lucovsky* [4.15] is the simplest that gives good agreement with experimental results. It gives σ_0 by the expression

$$\sigma_0 = \frac{2q^2 h}{3\pi nm^* c\varepsilon_0} \left(\frac{\mathscr{E}_e}{\mathscr{E}_0}\right)^2 \frac{\sqrt{E_A}(hv - E_A)^{3/2}}{(hv)^3}. \tag{4.19}$$

Upon substitution of numerical values, it becomes for Si

$$\sigma_0 \doteq 2.2 \times 10^{-17} \left(\frac{\mathscr{E}_e}{\mathscr{E}_0}\right)^2 \frac{m}{m^*} \lambda_c \left[\frac{\lambda}{\lambda_c}\left(1 - \frac{\lambda}{\lambda_c}\right)\right]^{3/2} [\mathrm{an}^2],$$

where $(\mathscr{E}_e/\mathscr{E}_0)$ is the effective electric field at the impurity center and $\lambda_c = 1.24/E_A$, given in μm. The functional dependence of σ_0 on wavelength is shown in Fig. 4.8. It rises from zero wavelength to a maximum at $\lambda_c/2$ and then decreases. Although it is not constant, it has a rather broad maximum and the absorption coefficient is reasonably constant over a useful wavelength range. For example, for Si:In, where $\lambda_c/2 = 4\,\mu m$, σ_0 is within 20% of its maximum value from 2.5 to 5.5 μm, clearly covering the 3–5 μm range.

The maximum value of σ_0 varies with energy level of the extrinsic impurity, as shown in Fig. 4.9. Note that the shallower the energy level, the larger the photoionization cross section.

Experimental verification of σ_0 is difficult to achieve. However, the detector responsivity is readily measured, according to the expression

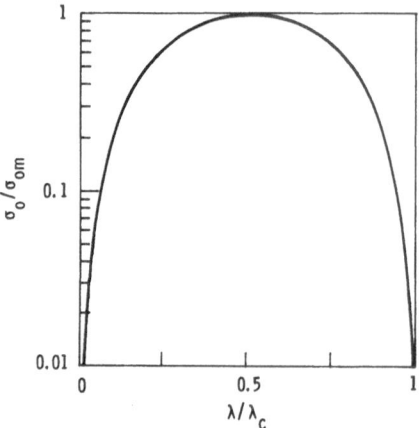

Fig. 4.8. Normalized photoionization cross section as a function of wavelength [4.15]

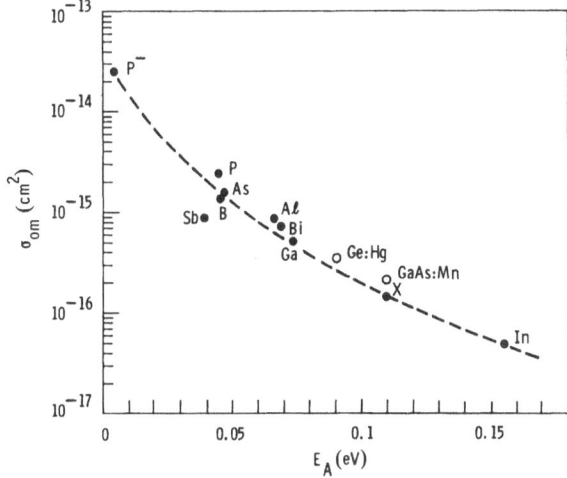

Fig. 4.9. Maximum value of photoionization cross section for various impurities [4.16]

$$R = I_{ph}/W ; \tag{4.20}$$

here I_{ph} is the detector photocurrent and W the incident power. The responsivity is also defined as

$$R = q\eta\lambda G/hc , \tag{4.21}$$

which, using (4.15) for $\alpha l < 1$, can be written as

$$R \simeq (1-R)q\alpha l\lambda G/hc , \tag{4.22}$$

i.e., $R \sim \alpha\lambda \sim \sigma_0\lambda$, the responsivity is proportional to σ_0. A plot of R vs λ is shown in Fig. 4.10 for Si:In. Both the experimental data and the theoretical curve of (4.19) are shown, and the agreement with the Lucovsky model is quite good.

It has been found by *Sclar* [4.10] that for a given energy level, n-type extrinsic impurities in Si have their peak response at longer wavelengths than do p-type impurities. Since the thermal carrier concentrations is determined by the energy level, n-type detectors are expected to provide superior temperature characteristics, for a given peak wavelength response.

The quantum efficiency expressions imply that high η values can be achieved by making the detector very thick. There is a limit to this in extrinsic detectors, however, because photocarriers generated beyond the drift length $l_d = \mu\tau\mathscr{E}$ recombine before being collected. This is shown in the photocurrent expression (4.4) by the gain G_0, which can be written as $G_0 = l_d/l$ and decreases as l increases. Increasing l to raise η becomes a self-defeating task once $l > l_d$. Fortunately, for most extrinsic detectors the drift length is sufficiently long that quantum efficiencies approaching 50 % can be obtained.

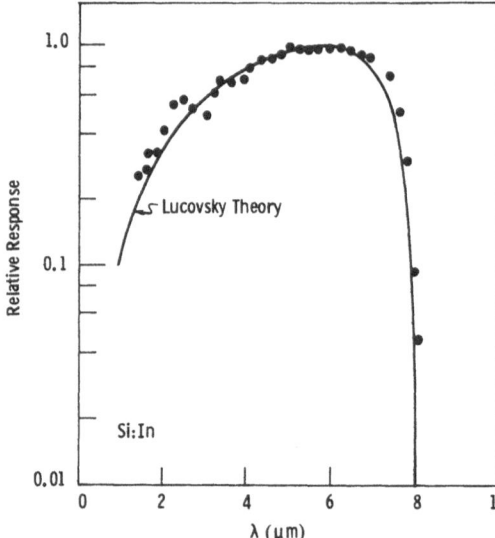

Fig. 4.10. Experimental (points) and theoretical spectral response of a Si:In detector [4.17]

4.2.5 Free Carrier Absorption

The quantum efficiency was defined as the ratio of the photon flux absorbed within the detector to the incident flux. This absorption is useful only if it takes place within the photosensitive portion of the device. For the detector configuration used in focal plane arrays, the incident flux must pass through the surface layer of the device, however, which is either heavily doped to facilitate contacting or has a CCD poly-Si gate structure on it. In either case, there are regions of heavily doped Si in which the carriers do not freeze out at the low detector operating temperature. As a result, there is a high concentration of free carriers, and photons can be absorbed by free carrier absorption, a mechanism in which free carriers are excited to higher energy states in their band and the resultant absorption coefficient is proportional to the free carrier concentration and the square of the wavelength. It does not result in a photoconductive signal and reduces the useful photon flux density in the active region. For high carrier concentrations, the free carrier coefficient can be as high as $1000\,\text{cm}^{-1}$, and in extreme cases, it can cause nearly $100\,\%$ absorption of the ir irradiation.

It is useful to write the transmissivity through a heavily doped layer in terms of the sheet resistance R_s of that layer and an average mobility $\bar{\mu}$ as

$$T \simeq (1 - R)\exp(-K\lambda^2/q\bar{\mu}R_s). \tag{4.23}$$

Schroder et al. [4.18] found that (4.23) describes the experimental data quite well, as seen in Fig. 4.11, for both p- and n-Si. For n-Si: $K/\bar{\mu} = 2.4 \times 10^{-20}$ and for p-Si: $K/\bar{\mu} = 5.4 \times 10^{-20}$, when λ is expressed in μm and R_s in ohms/square. Equation (4.23) allows the transmissivity to be expressed in terms of the easily measured variable R_s.

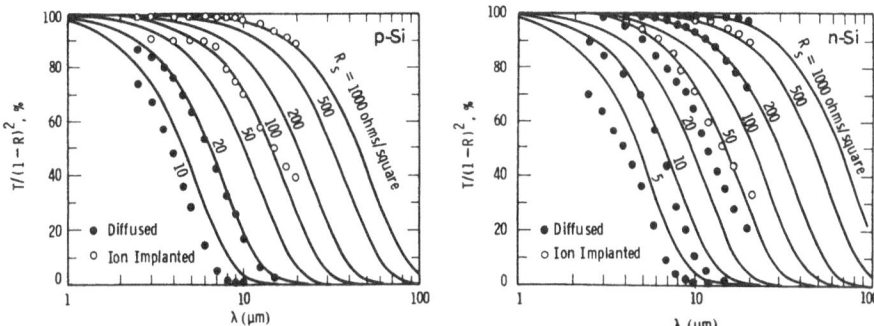

Fig. 4.11. Normalized transmissivity for diffused and ion-implanted layers. The curves were calculated from (4.23) and the experimental points in order of increasing transmissivity are for p-Si: 9, 25, 100, and 600 and for n-Si: 5.7, 12, 35, 56, 110, 360, and 400 ohms/square [4.18]

For $\gtrsim 90\%$ transmissivity, the sheet resistance of the p^+ or n^+ layers should be

p-Si: $R_s \lesssim 3\lambda^2$ ohms/square

n-Si: $R_s \lesssim \lambda^2$ ohms/square

with λ in units of micrometers. These values are easily achieved by conventional diffusion technology for the lower wavelengths, while for the 8–14 µm region ion implantation may have to be used.

4.2.6 Lifetime

The energy band diagram of Fig. 4.3 is repeated in simpler form in Fig. 4.12 to bring out the main points for lifetime considerations. For typical operating conditions, $p_{th} \ll p_{op} \ll N_D \ll N_A$. For Si:In in the 3.4–4.8 µm band, $p_{op} \simeq 10^6$ cm^{-3} for 300 K radiation, while $N_D \simeq 10^{13}$ and $N_A \simeq 1$–5×10^{17} cm^{-3}.

From Fig. 4.12 and the inequalities above, it is obvious that $N_A^- \simeq N_D$. Thus, the acceptor atoms being negatively charged by the compensating donors act as recombination sites for free holes, with the lifetime given by

$$\tau = (BN_D)^{-1}, \tag{4.24}$$

where B is the recombination or capture coefficient, which varies strongly with temperature, as shown in Fig. 4.13. Although there is a fair amount of scatter in the data taken from various papers, the dashed line gives a first-order analytical expression

$$B \simeq 1.5 \times 10^{-4}/T. \tag{4.25}$$

Note that the capture coefficient has similar values for both Si and Ge.

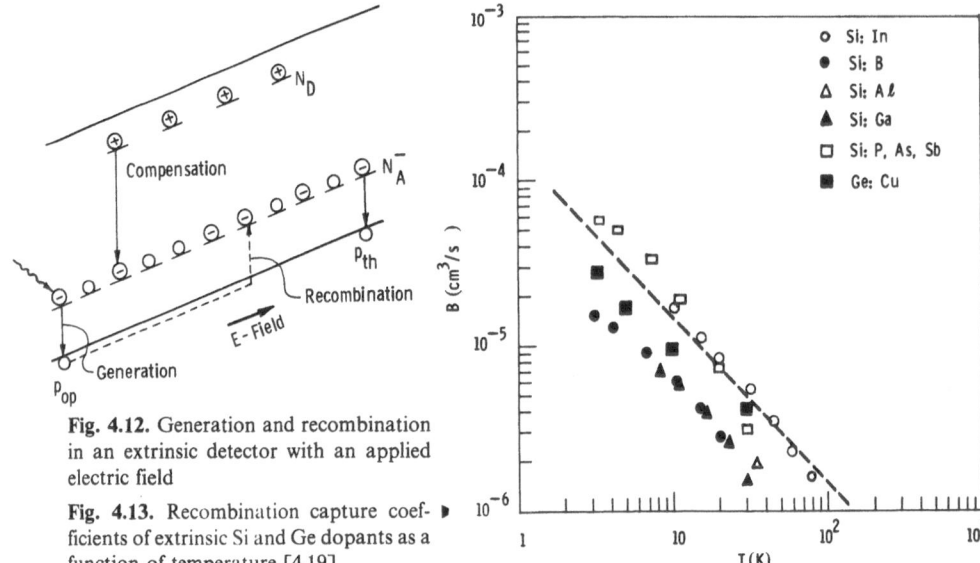

Fig. 4.12. Generation and recombination in an extrinsic detector with an applied electric field

Fig. 4.13. Recombination capture coefficients of extrinsic Si and Ge dopants as a function of temperature [4.19]

The lifetimes obtained from the B values of Fig. 4.13 apply at low electric fields. As the field across the device is raised, the carriers gain energy to increase their velocity and the probability of capture by coulombic attraction decreases. This causes the lifetime to increase with field, which can be expressed by the relationship

$$\tau \sim \mathscr{E}^n,$$

with $n = 0.5$ to 2 having been experimentally determined [4.10].

In p-Si with the inevitable boron contaminants, the compensated boron atoms can act as recombination centers. Whether they do depends on the temperature. For Si:In at 50–60 K, holes captured by boron impurities will be thermally emitted very rapidly and the B acts as a temporary trap, but not a recombination center. However, for Si:Ga, operating at 25–30 K, the thermal emission from B is much reduced and B does act as a recombination center. For this case, (4.24) holds, assuming the capture coefficients for Ga and B are equal. For Si:In the lifetime expression becomes $\tau = [B(N_D - N_B)]^{-1}$.

With these considerations and the previous discussion of p_{th}, the D^* expression of (4.8) can be written as

$$D^* = \frac{\eta\lambda}{2hc} [(lBN_A N_v/g) e^{-E_A/kT} + \eta Q_B]^{-1/2}, \qquad (4.26)$$

i.e., it is independent of lifetime. Nevertheless, the lifetime is important from photocurrent considerations, as we shall see later on.

It is clear from the discussion above that to increase the lifetime, the material should be closely compensated, i.e., $(N_D - N_B)$ should be small, but, consistent with p_{th}, as shown in Fig. 4.7, it should not be zero. It should typically be $\lesssim 10^{13}$ cm^{-3}. Now for Czochralski-grown Si, where $N_B = 5$–10×10^{13} cm^{-3}, it is obviously very difficult to achieve the desired compensation. For float zone, where the boron concentration is lower by a factor of 10–50, precise compensation is easier to obtain, provided a compensating impurity like phosphorus can be introduced at such low levels. This is a problem, because traditional doping methods, such as doping from the melt, are difficult to control at these low levels.

A promising new method is the use of neutron transmutation doping were the thermal neutrons in a nuclear reactor interact with the Si lattice to form phosphorus. A thermal neutron absorbed by a ^{30}Si isotope forms a ^{31}Si isotope. This in turn decays into a lower energy state through the emission of a gamma ray. Finally, the emission of a beta ray completes the transmutation step into ^{31}P.

The relationship between the final phosphorus concentration N_P, the neutron flux density, and the radiation time can be written as [4.20]

$$N_P = N_{Si}^{30} \sigma \Phi_n t, \qquad (4.27)$$

where N_{Si}^{30} is the concentration of ^{30}Si atoms/cm^3 (equal to 3.1 % of the total Si concentration), σ is the thermal neutron capture cross section (1.3×10^{-25} cm^2), Φ_n is the neutron flux density, and t is the radiation time. Substitution of numerical values gives

$$N_P = 2 \times 10^{-4} \Phi_n t. \qquad (4.28)$$

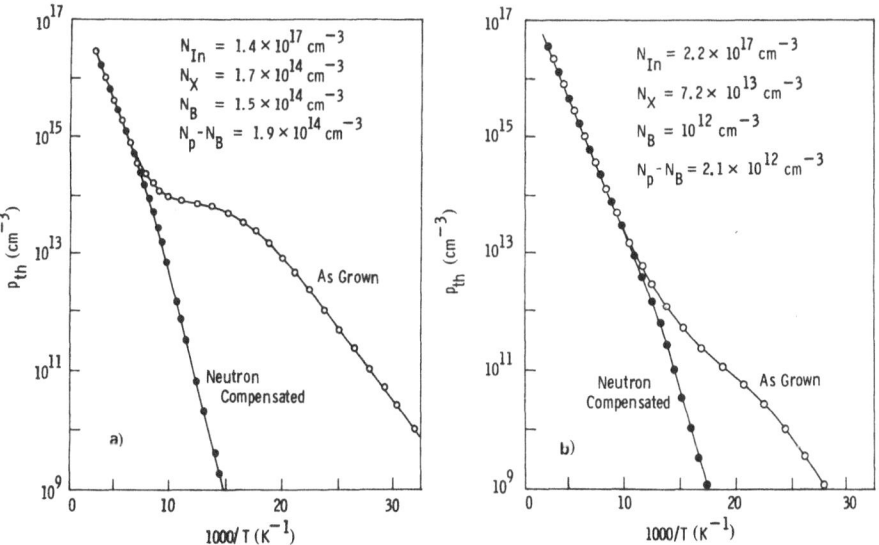

Fig. 4.14a, b. Carrier concentrations as a function of temperature (**a**) before and (**b**) after neutron transmutation doping. The concentrations on the figures refer to In, X (0.11 eV level), B, and P-B (net compensation) [4.17]

Typical reactor flux densities are around 10^{13} neutrons/s cm^2, leading to irradiation times of the order of an hour. The uniformity that has been obtained is $\leqslant 5\%$ and the doping concentration can be very precisely realized since the neutron flux density is well known in a given location in the reactor and the irradiation time can be precisely controlled.

Two examples of transmutation doping are shown in Fig. 4.14, where p_{th} is shown before and after appropriate P compensation. The Czochralski-grown sample had a fairly high N_B of 1.5×10^{14} cm^{-3} with a residual phosphorus concentration of 6×10^{13} cm^{-3}. After neutron irradiation $N_P - N_B = 2 \times 10^{14}$ cm^{-3}. The other sample, which was float-zone grown, had an initial N_B of 10^{12} cm^{-3} and a final $N_P - N_B = 2 \times 10^{12}$ cm^{-3}. These examples give a good demonstration of the power of the neutron transmutation doping technique for very precise compensation.

4.2.7 Mobility

The mobility is composed of three components, each the result of a different type of scattering mechanism. The first is lattice scattering which has been described by the equation [4.21]

$$\mu_{nl} \simeq 2 \times 10^9 / T^{2.5} ; \quad \mu_{pl} \simeq 2.3 \times 10^9 / T^{2.7} . \tag{4.29}$$

This is the dominant scattering mode for lowly doped Si at room temperature, where (4.29) gives the familiar 1300 and 480 cm^2/V s for n- and p-Si, respectively.

The second component is ionized impurity scattering, given by the expression [4.22]

$$\mu_i = \frac{5 \times 10^{17} T^{3/2}}{N_i \ln(1 + 4.5 \times 10^8 T^2 / N_i^{2/3})}, \tag{4.30}$$

where N_i is the ionized impurity concentration. For extrinsic detectors $N_i \simeq 2N_{comp}$, the compensating density, for $p_{op}, p_{th} \ll N_{comp}$. For typical compensation densities in the 10^{13}–10^{14} cm^{-3} range, μ_i contributes a negligible component to the mobility.

The third component is neutral impurity scattering, which has been given as [4.23]

$$\mu_n \simeq 1.24 \times 10^{22} E_A / N^0 , \tag{4.31}$$

where N^0 is the neutral impurity concentration. For $N^0 = 10^{16}$–10^{17} cm^{-3}, typical for extrinsic detectors, neutral impurity scattering is the dominant component at low temperatures.

The overall mobility is

$$\mu = (1/\mu_l + 1/\mu_i + 1/\mu_n)^{-1}. \tag{4.32}$$

Since μ_i can usually be neglected, we have found that an approximate expression for the hole Hall mobility for Si:In is

$$\mu \simeq \frac{2.5 \times 10^{21}}{N^0 + 7 \times 10^{11} T^{2.85}}. \tag{4.33}$$

This equation gives fairly good agreement with the measured Hall mobility over the 20–300 K temperature range, provided ionized impurity scattering does not contribute significantly. For $N^0 \simeq 10^{17} \, \text{cm}^{-3}$, this means that $N_i \lesssim 10^{14} \, \text{cm}^{-3}$. If it is higher, then (4.32) should be used.

Typical values of Hall mobility for Si:In taken at low electric fields are in the $0.5–1 \times 10^4 \, \text{cm}^2/\text{V s}$ range at $T \simeq 50 \, \text{K}$. In a detector, the fields are much higher than those used for Hall measurements and there is a mobility decrease with electric field increase. From being independent of field, the mobility becomes $\mu \sim 1/\sqrt{\mathscr{E}}$ and eventually $\mu \sim 1/\mathscr{E}$, making the velocity, $v = \mu\mathscr{E}$, independent of the field. For extrinsic Ge the $\mu \sim 1/\sqrt{\mathscr{E}}$ relationship begins at fields of 10–100 V/cm, while for Si it happens at 100–1000 V/cm [4.24]. For detector fields of around 1000 V/cm, the mobility calculated from (4.33) may not be true, the real value being lower. A further caution is that the Hall mobility of (4.33) is equal to the drift mobility only when the Hall factor equals unity. This is approximately correct when neutral impurity scattering is dominant, the case for extrinsic detectors at low temperatures.

4.2.8 Cross Talk

The modulation transfer function of the MFPA is degraded by cross talk, be it electrical or optical in the imaging array or in the associated readout device. If the readout device is a CCD, there is some cross talk introduced due to nonperfect charge transfer. With charge in the well of every cell, after N cell transfers, the cross talk is $N\alpha/(1-\alpha) \simeq N\alpha$ [4.25], where α is the fractional loss per cell. For $N = 50$ and $\alpha = 5 \times 10^{-4}$, it is 2.5%. If an isolation cell is placed between adjacent charge packets, then the expression becomes $N(N-1)\alpha^2/[2(1-\alpha)^2] \simeq (N\alpha)^2/2$. For the example above, this gives 0.03% and shows that by providing isolation bits, cross talk can be significantly reduced. This has been incorporated into the design of some MFPAs.

The other source of cross talk is the detector array itself. Electrical cross talk is generally negligible because carriers move by drift and the electric field keeps them well confined to the actual detector dimensions in the array. Optical cross talk tends to be more serious because photons enter the detector at some angle

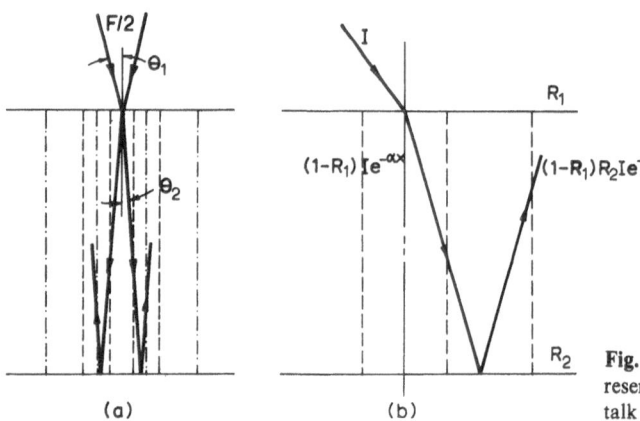

Fig. 4.15a, b. Schematic representation of optical cross talk

Θ_1, determined by the f/number of the optics. However, as a result of the higher refractive index of Si, this incident angle is reduced to Θ_2 indicated on Fig. 4.15. The two angles are related by

$$\sin\Theta_2 = (1/n)\sin\Theta_1 \simeq 1/2nF,$$

where n is the refractive index and F is the f/number.

The schematic in Fig. 4.15a is drawn to scale for a substrate 500 μm thick with detectors on 50 μm (dashed lines) and 100 μm (dot-dash lines) centers. Note that for $F/2$ optics, the rays incident on the center of the detector spill over into adjacent detectors for the smaller elements, but are confined to the larger ones by the time they reach the bottom surface. Nothing can be done about this type of cross talk, except to keep the f/number high to reduce the angle of incidence. The intensity of the irradiation is of course reduced as it traverses the detector due to absorption. Typically, however, $\alpha \simeq 10\,\mathrm{cm}^{-1}$ so that for a thickness of 500 μm, $1 - \exp(-\alpha l) = 0.4$, i.e., only about 40% of the photons are absorbed by the time the bottom surface is reached. Photons reflected from that surface will further contribute to cross talk, as shown in more detail in Fig. 4.15b. This figure also shows the importance of the bottom surface reflectivity R_2. Obviously, for large detectors and large f/numbers, it is desirable to keep R_2 as close to unity as possible, since for this case the reflected rays remain within a detector element and additional passes generate more photocarriers. However, for the example of Fig. 4.15b, R_2 should be made negligibly low to prevent the spread of incident irradiation into adjacent elements.

Optical cross talk at the present stage of extrinsic Si development is around 5% for detector sizes that are consistent with high-density two-dimensional MFPA arrays [4.26]. Techniques to reduce this effect include back surface treatment to reduce R_2, increase the acceptor doping concentration to increase α, and means to provide some type of optical shields on the light-admitting side of the array.

4.3 Long Wavelength Detectors

There are basically two approaches to long wavelength extrinsic imaging. In the first, impurities with very shallow energy levels are utilized in the conventional manner. This works for Ge out to wavelengths of around 120 μm with B and Ga and to 300 μm for n-GaAs (see Fig. 4.2). For Si, however, shallow impurities like B, P, and Sb have energies around 0.04 eV giving cutoff wavelengths around 30 μm. This would indicate that extrinsic Si cannot be used for long wavelength imaging.

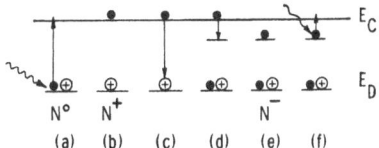

<table>
<tr><td>N^0</td><td>N^+</td><td></td><td></td><td>N^-</td><td></td></tr>
<tr><td>(a)</td><td>(b)</td><td>(c)</td><td>(d)</td><td>(e)</td><td>(f)</td></tr>
</table>

Fig. 4.16a–f. Energy band diagram of far infrared photodetection using excited states of Si: P

The second approach is that proposed by *Norton* [4.27]. Here P-doped Si is used as shown in Fig. 4.16. Background radiation, such as 300 K radiation, excites electrons from the neutral atoms N^0 into the conduction band as shown by (a). The free electron in (b) has two choices: it can either recombine with the ionized donor as in (c) or be captured onto an excited level of a neutral atom (d), making it negatively charged (e). The relatively weak binding energy of the extra electron makes the N^- center useful as a detector for wavelengths out to 200 μm. Such a long wavelength response means that the binding energy is only about 0.004 eV. This, of course, also implies that for negligible thermal excitation, the operating temperature must be very low, typically 2–4 K.

Initial experiments [4.27] indicate that for P doping concentrations of around 10^{16} cm^{-3}, approximately 10^{13} cm^{-3} are in the N^- state. This is a fairly low percentage. But since the photoionization cross section is quite large (see Fig. 4.9), useful quantum efficiencies can be obtained. Values of 1–5 % have been measured.

4.4 Readout Techniques

There are basically two methods to read out the photogenerated charge in extrinsic MFPAs. In the first, shown in Fig. 4.17a, the extrinsic substrate alone is used. By virtue of the low operating temperature, the substrate can be considered an insulator and the applied gate voltage divides between it and the gate oxide. Holes generated by absorbed photons drift to the oxide/Si interface to be transported along that interface as *majority* carriers. The device was first described in 1974 [4.28], but has not found application for focal plane arrays, *partly because it has poor transfer efficiency.*

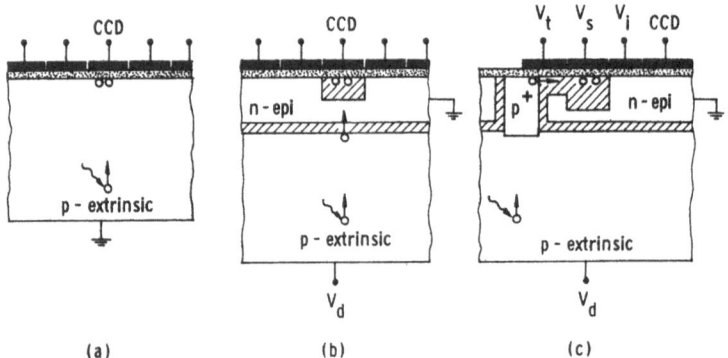

Fig. 4.17. (a) Majority and (b), (c) minority carrier CCD readout techniques

The second approach, shown in Fig. 4.17b, c, consists of the extrinsic substrate and a layer of the opposite conductivity type. Majority carriers in the extrinsic substrate are injected into the upper epitaxial layer where they become minority carriers. The CCD is a *minority* carrier device, operated in the conventional manner.

The two injection schemes shown in Fig. 4.17b, c are basically different from one another. The operation of Fig. 4.17b is similar to a bipolar transistor. Photocharge from the extrinsic substrate (emitter) is injected into the signal processing epitaxial layer (base) to be collected by the gate induced potential well (collector). Bipolar transistor considerations must be taken into account. For example, recombination in the base and the emitter-base space-charge region reduces the collection efficiency. Space-charge region (scr) recombination is particularly important at the low currents characteristic of infrared detectors, particularly when operating in the 3–5 μm wavelength band. To eliminate both scr and neutral base recombination, the gate voltage can be made sufficiently high to cause the gate-induced potential well to punch through to the emitter. Then the holes from the photoconductor are never in contact with majority electrons in the epitaxial layer and no recombination takes place. In contrast to punch-through in a conventional bipolar transistor, where a catastrophic current increase occurs, in an MFPA the photoconductor can be considered a constant current generator and nothing drastic happens when punch-through takes place.

The injection scheme of Fig. 4.17c can be compared to an MOS field effect transistor. A diffused region (source) connects the signal processing epitaxial layer with the extrinsic photoconductor. The transfer gate V_t (gate) allows the photocharge to flow from the source into the potential well under the storage gate V_s (drain). This mode of injection is generally called direct injection, in spite of the fact that the injection of Fig. 4.17b appears to be more direct. The input circuit can be considered a grounded gate MOSFET, in which the source biases itself to a voltage to exactly accommodate the photocurrent from the extrinsic

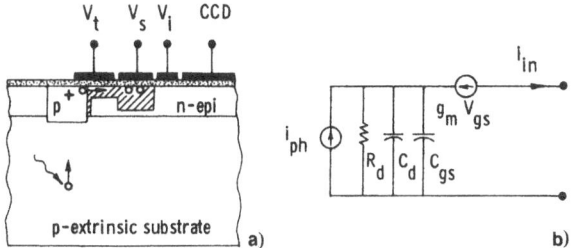

Fig. 4.18a, b. Direct injection configuration of Fig. 4.17c; (a) device, (b) equivalent circuit

substrate. This mode of operation, being current controlled, is threshold voltage insensitive, an important advantage for imaging arrays in which threshold voltage variations are typically around 100 mV. The direct injection input has been analyzed in detail and the theory verified by experiment [4.29, 30].

A third injection method is the gate modulation technique [4.30] (Chap. 5). This method is generally not used for direct-coupled extrinsic MFPAs, but has been used for coupling HgCdTe photoconductors to Si CCDs [4.31].

4.4.1 Direct Injection

The direct injection circuit of Fig. 4.17c is redrawn in Fig. 4.18, together with its equivalent circuit. In the equivalent circuit, R_d is the detector resistance, C_d its capacitance, C_{gs} the effective gate/source capacitance of the input MOSFET, for which the p^+ diffusion is the source, "V_i" the gate, and "V_s" the drain. The input current is the photocurrent i_{ph} and the MOSFET is characterized by a transconductance g_m.

The current flow i_{in} into the CCD is

$$i_{in} = \frac{g_m}{(G_d + g_m)} \frac{i_{ph}}{[1 + (f/f_0)^2]^{1/2}}, \qquad (4.34)$$

where $f_0 = (G_d + g_m)/2\pi C$ is the direct injection frequency, with $C = C_d + C_{gs}$ and $G_d = 1/R_d$. Equation (4.34) shows that the injection frequency is limited by f_0, which, as we shall see later, is determined by the photon flux density.

The aim of the direct injection circuit is to inject all of the photocurrent into the CCD. Two leakage paths exist to try to prevent this from happening. One is via the detector resistance and to minimize this leakage $G_d \ll g_m$. This condition is generally met in extrinsic Si devices because their resistance is very high. A note in passing: this condition is not met in the hybrid HgCdTe photoconductor/Si CCD because the detector resistance is very low and a buffer amplifier is therefore required. Direct injection, however, is possible in hybrid configurations, when the detector is a diode with high junction resistance [4.30].

The second leakage path is via capacitance C. However, this is effective only for frequencies above f_0. To reduce this leakage path, it is important to make C as small as possible. Again extrinsic Si fares well, because C can be made of the order of 0.1–0.5 pF. This is not true for such hybrid combinations as PbSnTe diodes/Si CCD, where the diode capacitance is much higher. The alternate way of increasing f_0 by increasing g_m is not possible since the transconductance is determined by the photocurrent and cannot be increased through circuit layout, as shown in the following section.

Transconductance

In contrast to conventional MOSFETs that operate at high currents, the equivalent MOSFET of the direct injection circuit operates in the subthreshold mode. In this mode the transconductance is given by [4.32]

$$g_m = qi_{in}/nkT, \tag{4.35}$$

i.e., it is determined by the photocurrent alone, not by the device parameters. The factor n accounts for interface states at the Si/SiO$_2$ interface and fixed charge in the space-charge region under the gate. Its value is typically 1.5–2. Equation (4.35) holds very well for devices sensitive in the 3–5 µm band, where the photon flux density is low. For the 8–12 µm band the photocurrent is higher and the MOSFET operation is driven closer to the above-threshold condition.

A transconductance expression that is valid for both regimes of operation is [4.29]

$$g_m = n\beta \frac{kT}{q} \left[\sqrt{1 + \frac{2i_{in}}{\beta(nkT/q)^2}} - 1 \right], \tag{4.36}$$

where $\beta = Z\mu C_0/L$ and Z/L is the width/length ratio of the transfer gate and C_0 is the gate capacitance/unit area. Equation (4.36) reduces to (4.35) in the low current limit. An experimental verification of (4.36) is shown in Fig. 4.19, where g_m was measured over a wide current range and the agreement of theory and experiment is quite good over the 27–300 K temperature range. The increasing n value at lower temperatures is probably due to the higher interface state density near the band edge and the increased β is due to a higher field effect mobility. In fact, the mobility increases by about a factor of 7 in going from 296 to 27 K. The photocurrents in the 3.4–4.8 µm wavelength band are around 10^{-10} amps for a 10^{-4} cm^2 area detector. For the 8–12 µm band the current is approximately a hundred times higher.

The transconductance can be increased by injecting a dc bias current in addition to the photocurrent. This has two disadvantages, however. First, the bias current acts like a background current reducing the storage well capacity and second it introduces shot noise, making it difficult or impossible to achieve BLIP performance.

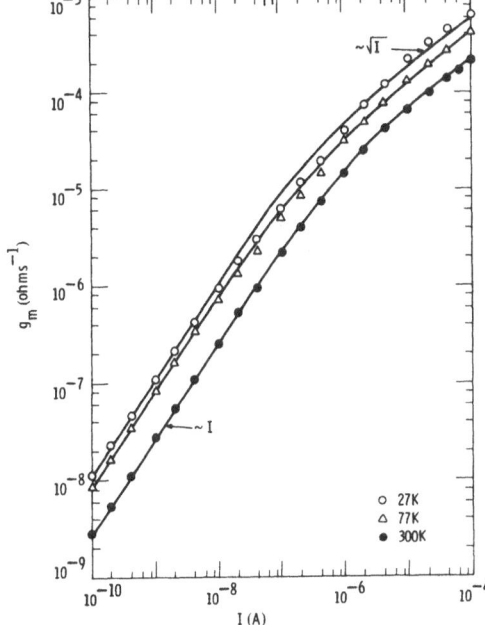

Fig. 4.19. Transconductance of a Si MOSFET as a function of current and temperature. The solid lines are calculated from (4.36) with $n=1.4$, $\beta=2.4 \times 10^{-4}$ at 300 K, $n=1.85$, $\beta=9\times10^{-4}$ at 77 K and $n=3.65$, $\beta=1.8\times10^{-3}$ at at 27 K, and the points are experimental data

Noise

In an MFPA there are some noise sources that are not present in single detector/amplifier systems. On the other hand the use of low noise CCDs for multiplexing and preamplifiers, both of which are on-chip, reduces the noise compared to discrete systems. Nevertheless, to achieve BLIP behavior, very stringent requirements are placed on the design and fabrication of the array.

The direct injection method will be used for the noise analysis here, because extensive work – both theoretical and experimental – has been done for this mode of injection.

The noise sources in an extrinsic Si MFPA are shown in Fig. 4.20. The analytic formulation is given in terms of noise currents and is similar to that of

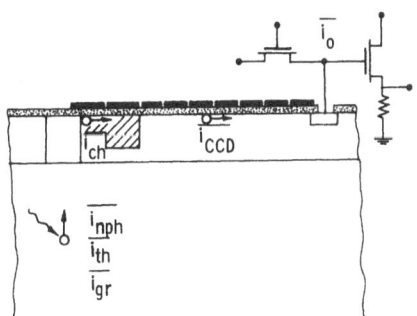

Fig. 4.20. Noise sources in a direct injection MFPA

[4.29], with the exception of dealing with a photoconductor with a frequency-dependent gain.

We assume in the following that the array is operated in the time delay and integration (TDI) mode with N detectors in the TDI direction. The noise sources are:

a) Photon noise, due to the random nature of the absorbed irradiation

$$\overline{i_{nph}^2} = 4q^2 \eta N Q_B A \mathscr{F} |F_1(f)|^2 \; A^2/Hz, \tag{4.37}$$

where

$$\mathscr{F} = [g_m/(g_m + G_d)]^2 \; sinc^2(\gamma f/f_c)$$
$$|F_1(f)|^2 = \sum_{n=-\infty}^{\infty} U_1(f + nf_c)$$
$$U_1(f) = G^2 \; sinc^2(f/f_c)/[1 + (f/f_0)^2]$$
$$sinc(x) = \sin(\pi x)/(\pi x)$$

and the frequency dependent gain is taken to be [4.11]

$$G^2 = G_0^2[1 + (f/2G_0 f_e)^2]/[1 + (f/f_e)^2],$$

where γ denotes the fraction of the clock period during which the reset gate at the output is off, and f_c is the clock frequency.

The sinc functions account for the sampled nature of the input circuit and the output sample-and-hold.

For the case of $g_m \gg G_d$ and neglecting the sampling effects, (4.37) simplifies to

$$\overline{i_{nph}^2} = 4q^2 \eta N G^2 Q_B A. \tag{4.37a}$$

b) Detector noise, due to thermal and generation-recombination fluctuations

$$\overline{i_{th}^2} = (4kTN/R_d)\mathscr{F}|F_2(f)|^2, \tag{4.38}$$
$$\overline{i_{gr}^2} = (4q^2 G^2 p_{th} NlA/\tau)\mathscr{F}|F_2(f)|^2, \tag{4.39}$$

where $F_2(f)$ is similar to $F_1(f)$ except that it does not contain the gain G. The simplified expressions for (4.38) and (4.39) are

$$\overline{i_{th}^2} = 4kTN/R_d, \tag{4.38a}$$
$$\overline{i_{gr}^2} = 4q^2 G^2 p_{th} NlA/\tau. \tag{4.39a}$$

c) Channel noise of the direct injection equivalent MOSFET consists of thermal and $1/f$ noise. For simplicity only thermal noise is considered here, i.e.

$$\overline{i_{ch}^2}=4kTNg_m\mathscr{F}\left[|F_2(f)|^2/(g_mR_d)^2\right.$$
$$\left.+(f_c/\pi f_0)|F_3(f)|^2\right],\qquad (4.40)$$

where

$$|F_3(f)|^2=\sum_{n=-\infty}^{\infty}U_3(f+nf_c)$$
$$U_3(f)=\sin^2(\pi f/f_c)/[1+(f/f_0)^2].$$

For the simplified case, this becomes

$$\overline{i_{ch}^2}=4kTNg_m(f/f_0)^2.\qquad (4.40a)$$

d) CCD noise due to interaction of the charge packets with interface states,

$$\overline{i_{CCD}^2}=2.8\,q^2kTN_{st}A_bN_bf_c[1-\cos(2\pi f/f_c)]\,\text{sinc}^2(\gamma f/f_c),\qquad (4.41)$$

where A_b is the CCD gate area, N_b is the number of transfers, and N_{st} is the interface state density. The dark current noise of the CCD can be neglected due to the low temperature.

e) Output noise, due to noise in the on-chip electrometer amplifier, referred to the CCD output node,

$$\overline{i_0^2}=f_c^2C_{op}^2(4kT/g_{mo}+K_2/f),\qquad (4.42)$$

where C_{op} is the capacitance of the output node and g_{mo} is the transconductance of the amplifier, which also has some $1/f$ noise. The "kTC" noise which results from resetting the output integrator is not included, since it can be removed by correlated double sampling [4.33].

By cooling the detector sufficiently, p_{th} can be reduced to make the g–r noise insignificant. This requirement establishes the temperature of operation and it is obvious that the lower the background flux density the lower must be the operating temperature. The channel noise is partly determined by the photocurrent [see (4.35)], but its $1/f$ component is influenced by processing. CCD noise can be reduced by decreasing the interface state density, and the output noise is strongly influenced by the output node capacitance.

The responsivity is

$$R=q\lambda\eta GN/hc\qquad (4.43)$$

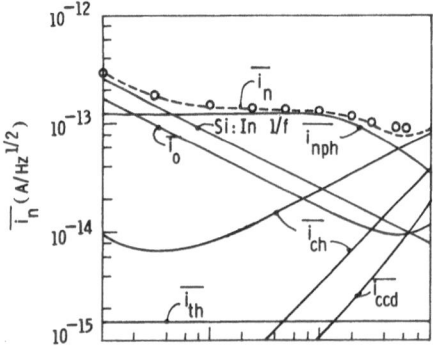

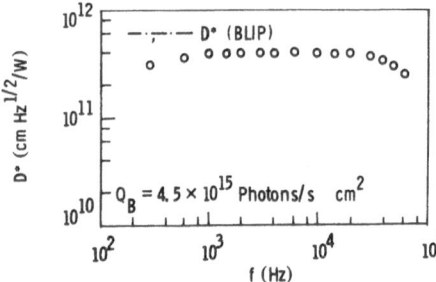

Fig. 4.21. Noise currents and D^* for a 16-element Si:In TDI array. $Q_B = 4.5 \times 10^{15}$ photons/s cm^2, $T = 40$ K [4.34]

and the detectivity is defined by

$$D^* = R\sqrt{A}/i_n,\tag{4.44}$$

where i_n is the total noise current obtained by the quadrature addition of the noise currents (4.37–42). For photon noise limited performance, (4.44) reduces to the familiar form

$$D^* = \frac{\lambda}{2hc}\sqrt{\frac{\eta N}{Q_B}},\tag{4.45}$$

which differs from the single detector expression by $\sqrt{N}$, i.e., the D^* improves as a result of the signal addition in the TDI direction.

Calculated noise currents and experimental data points for a 16-element array of Si:In detectors direct injected, taken from [4.34], are shown in Fig. 4.21. The noise currents have the shape predicted by theory, and below about 1 kHz, $1/f$ noise dominates. Beyond 25 kHz direct injection frequency limitations cause a D^* roll-off. These data show that even at this early stage of extrinsic Si MFPA development, BLIP behavior is almost attained.

4.4.2 CCD Considerations

The CCD requirements for an MFPA are: low noise, high transfer efficiency, low cross talk, high-speed operation, and the ability to function at the low focal plane temperatures. The noise is mainly determined by interface states which can be reduced to acceptably low values for devices fabricated on properly annealed (100) oriented wafers. The noise due to CCD dark current is negligible since it is extremely low at the operating temperature. The transfer efficiency is also governed by interface states at low frequencies, typically less than 0.5–1 MHz. Transfer inefficiencies of around 10^{-4} per transfer are achieved and no significant change has been noted for operating temperatures as low as 10 K. Frequently, there is a small increase in transfer inefficiency at 25–30 K, the temperature where the electrons in the epitaxial layer begin to freeze out; however, in general the transfer efficiency is good enough.

The transfer inefficiency at high frequencies is largely determined by the gate length and field effect mobility. Since mobility increases with decreasing temperature (see Fig. 4.19), the operational frequency should increase. Frequencies up to 10 MHz are predicted for short gate ($\simeq 7\,\mu$m) p-surface channel devices. Higher frequencies will be difficult to achieve with surface channel devices, especially since at present most extrinsic impurities are p-type (e.g., Ga, In, and Al), requiring p-channel CCDs. n-type impurities have the advantage of being able to use n-channel devices with their increased mobility. Buried channel devices can achieve higher frequencies and may be required in the future.

The cross talk problem has been discussed earlier and does not appear to present major difficulties, especially if isolation bits are provided. The very fact that CCDs even function essentially as well at the very low temperatures as they do at room temperature is interesting. Remember at 10–20 K even such shallow level impurities as phosphorus are not thermally ionized. Nevertheless, they do work and make the implementation of extrinsic Si MFPAs possible.

4.5 Some MFPA Designs

The very significant effort presently being expended on extrinsic Si MFPAs rests on several promises of this technology:

1) perform a number of signal processing functions on the focal plane, e.g., preamplification, TDI, multiplexing, filtering, and even post-amplification,

2) reduce the cost per detector,

3) provide very high detector densities with the resultant improved performance, i.e., higher sensitivity or lower minimum resolvable temperature.

The basic approaches to ir imaging are discussed in Chap. 5. For extrinsic arrays two basic approaches are being pursued; one is the TDI mode and the other is the staring mode. In the TDI approach, two options are available, shown schematically in Fig. 4.22a, b. Both have the TDI advantages, viz. slow-

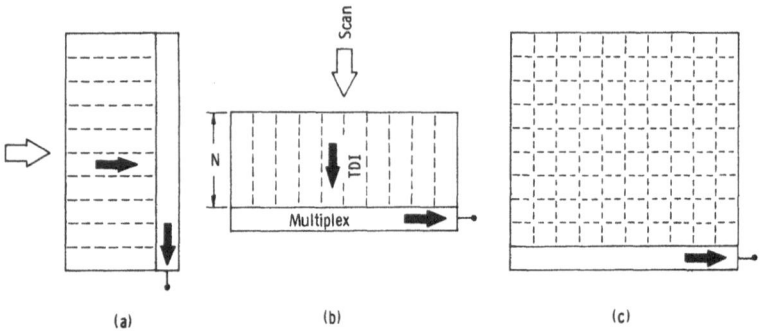

Fig. 4.22a–c. Basic MFPA layouts showing (**a**) azimuth, (**b**) elevation-scanned TDI arrays and (**c**) a staring array

speed scan mirror, uniformity, and signal/noise ratio increase due to the addition of the signal. The azimuthal scan approach is similar to present single-line parallel scan systems. It has the disadvantage that the multiplexed information gives vertical lines, which are noncompatible with conventional television display systems. To overcome this problem the output signals can be fed to a light-emitting diode array, as in some present parallel-scanned systems [4.35], a scan converter can be used to convert from vertical to horizontal lines or a vertically scanned display can be utilized.

None of these problems exists in the elevation-scanned system of Fig. 4.22b. The multiplexer (MUX) gives horizontal lines that are directly compatible with conventionally scanned TV displays. Another problem that is automatically solved in this implementation is the temperature step at the horizon, because the scan is across the horizon, not along it. It does place the onus of handling the resultant large dynamic range on the display, however. Although elevation scan systems have not generally been used in the past, they are natural for MFPAs which have both TDI and MUX on the same chip.

The ultimate imager is the staring array of Fig. 4.22c. Here a parallel-serial readout format is used, similar to the TDI arrays, but it requires no scene scanning. The readout is similar to the interline transfer format used in visible imaging arrays. Frame transfer techniques are rarely used for MFPAs, because of the large area on the cold finger which is not optically active. Staring arrays have traditionally not been used in ir imaging (except for the pyroelectric ir vidicon) because the uniformity has not been sufficiently good. Since uniformities are not likely to improve to the required few percent level in the near future, these arrays will be initially used in surveillance systems, where frame subtraction allows some nonuniformities. The frame subtraction approach may find future applications in terrestrial ir imaging, where through the use of semiconductor memories, the nonuniformities are removed electronically. Alternatively, *ac* coupling on a per-pixel basis may be a viable technique to remove the *dc* background and overcome some nonuniformities. This is especially important when the *dc* background charge is sufficiently high to saturate the CCD potentials wells during an integration time.

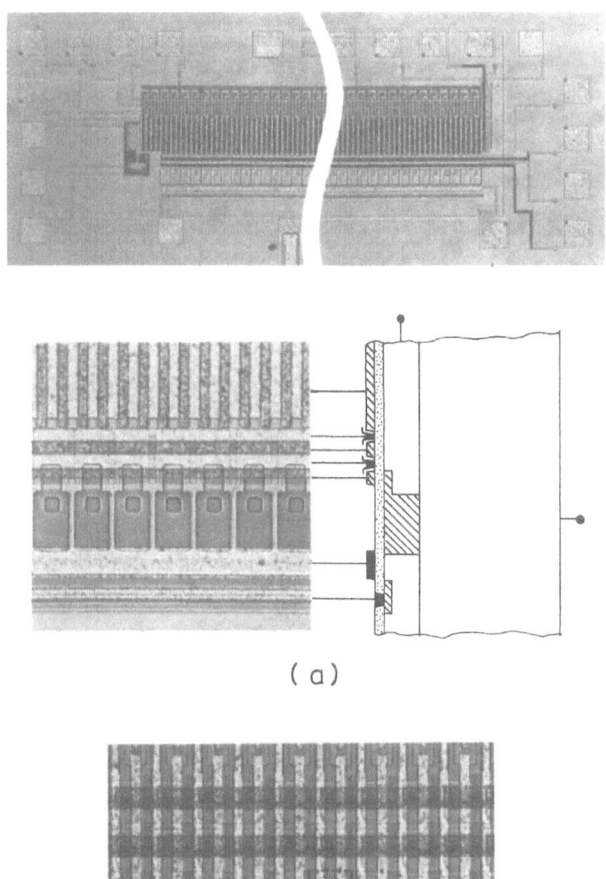

(a)

(b)

Fig. 4.23a, b. Si:In monolithic focal plane arrays; (a) 32-element line array, (b) part of a 4 × 32 area array (courtesy Westinghouse Electric)

Although much of the extrinsic work is proprietary or classified, a few layout examples will be shown. Both single line and area arrays have been fabricated for both modes. A photograph of a 32-element line array is shown in Fig. 4.23. The detectors are located beside the CCD and the photocurrent is injected via a series of input gates, as shown more clearly in the lower portion of Fig. 4.23a. The CCD is a four-phase device and the detectors are $25\,\mu m \times 50\,\mu m$. The responsivity varies typically $\pm 15\,\%$ along the 32 elements, which is similar to results on other arrays that have been reported.

A portion of a 4×32 area array is shown in Fig. 4.23b. Here the detectors lie under the CCD and the unit cell is only $40\,\mu m \times 45\,\mu m$. This is the sort of *detector size* envisioned for future FLIR (forward looking infrared) systems.

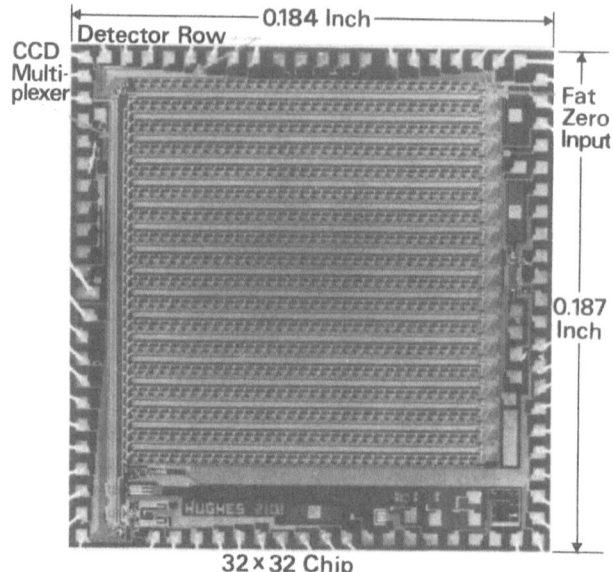

Fig. 4.24. 32×32 monolithic focal plane array (courtesy Hughes Aircraft)

An example of a 32×32 staring array is shown in Fig. 4.24. Its unit cell is $100\,\mu\text{m} \times 100\,\mu\text{m}$. The array has a bias charge input circuit on one side to improve the transfer efficiency of the imaging section CCD. At the other side is the MUX CCD which reads the image into an on-chip amplifier.

4.6 Summary

The development of extrinsic Si MFPAs is only a few years old, but already significant progress has been made during that time. Of course, the technology of previously developed Si CCDs and materials growth were there to be drawn upon.

4.6.1 Materials

Both Czochralski (CZ) and float-zone (FZ) techniques are used for crystal growth. CZ-grown Si has higher concentrations of boron and oxygen. The use of high-purity crucibles, however, has made it possible to reduce the boron concentration to the $10^{13}\,\text{cm}^{-3}$ level. FZ techniques produce similar crystals with appreciably less boron and oxygen, but they are somewhat more difficult to grow. The heaviest emphasis has been on Si:In and Si:Ga with In concentrations in the mid $10^{17}\,\text{cm}^{-3}$ and Ga typically in the high $10^{16}\,\text{cm}^{-3}$ range. An unknown level at 0.11 eV above the valence band in Si:In [4.16, 17, 36] and 0.056 eV in Si:Ga [4.36] have been found whose origin appears to be an indium-carbon pair [4.37]. It degrades device performance by either requiring excess cooling or, if properly compensated, by a decreased lifetime.

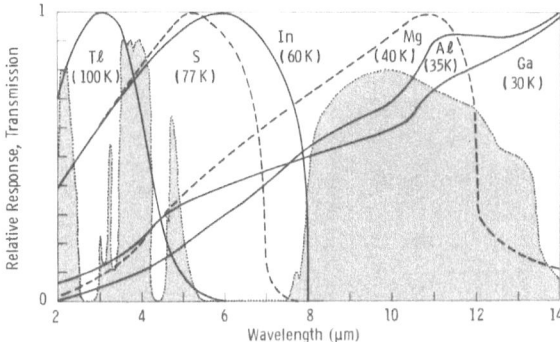

Fig. 4.25. Relative atmospheric transmission (shaded areas) and the relative response of several extrinsic Si detectors. The solid lines are for *p*-type and the dashed lines for *n*-type impurities. Approximate upper operating temperatures are shown in brackets (detector responses after SCLAR [4.10]

4.6.2 CCDs

The operation of CCDs, even when fabricated on epitaxial layers grown on extrinsic substrates, appears to present no problems. They function well to temperatures as low as 10 K with no significant deterioration. In fact, frequently noise and transfer efficiency improve, making them adequate for MFPAs.

4.6.3 Detectors

The important wavelength ranges are 3 to 5 and 8 to 14 μm, with some interest in the 2 to 2.7 μm band, as shown in Fig. 4.25. The main impurities used for the first two windows are In and Ga. Both, however, have cutoff wavelengths that are too high and therefore require excessive cooling. A number of other impurities have been investigated and a few are shown on Fig. 4.25. Sulfur is more suitable from a temperature point of view. However, its solubility is lower than In and it diffuses faster, leading to possible contamination of the epitaxial layer. Thallium is suitable because it has a low diffusion coefficient. Its solubility has not been established yet, however, and it does have a short cutoff wavelength. However, it is suitable for the 3.4–4.2 μm window.

For the longer wavelength detectors, Ga is chiefly used. Magnesium has a more suitable energy level, but according to SCLAR [4.10] appears to have a shallow 0.044 eV level associated with it, requiring low temperatures or extra compensation. If this level can be eliminated, then perhaps Mg is the ideal 8 to 14 μm level. For 3 to 5 μm an *n*-type impurity at around 0.2 eV would be ideal. It would raise the operating temperature to around 77 K, allow *n*-channel CCDs to be used, and could utilize the ubiquitous boron as a compensating impurity. This ideal level has yet to be discovered, however.

The use of neutron transmutation doping to introduce compensating phosphorus impurities into *p*-type substrates is a powerful technique. It has been shown [4.38] that for float-zone grown Si:In, responsivities as high as 100 A/W can be achieved and lifetimes as high as 200 ns have been measured. Such high-quality detectors should result in very high performance MFPAs.

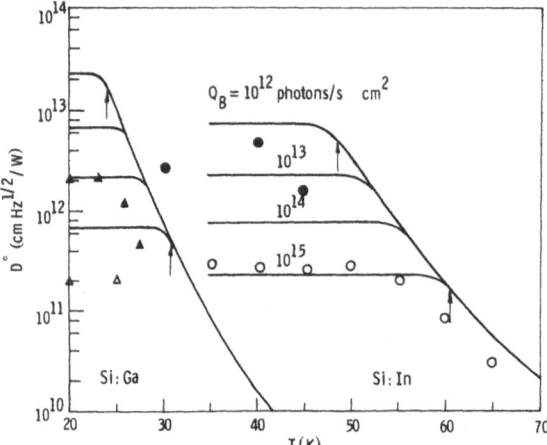

Fig. 4.26. Extrinsic Si single detector D^* values. The solid lines are calculated using (4.25) and (4.26) with $\eta = 50\%$ and $l = 5 \times 10^{-2}$ cm at $\lambda = 12\,\mu$m (Si:Ga) and $4\,\mu$m (Si:In). The data points are measured values (courtesy Hughes Aircraft and Westinghouse Electric). The circles are for Si:In (open circles: $Q_B = 10^{15}$, solid: 1.4×10^{12}) and the triangles are for Si:Ga (open: 9×10^{15}, solid: 2.4×10^{12} photons/s cm^2)

The D^* behavior of some Si:Ga and Si:In detectors is shown in Fig. 4.26 as a function of temperature. In general, the Si:In data lie 5–10 K below theory, mainly as a result of the 0.11 eV level contaminant, while the Si:Ga data are about 3 to 5 K below. As expected, the temperature decreases for reduced backgrounds, as shown by the 3 dB arrows. For example, for Si:In, T goes from 60 K for 10^{15} to 50 K for 10^{12} photons/s cm^2.

4.6.4 MFPAs

All of the material, CCD, and detector considerations are very important for MFPAs. But, more than that, everything must work properly in the MFPA format, where additional complications, such as injection, have to be considered. However, since we are dealing exclusively with Si, where the state of the art is highly developed, the design, understanding, layout, and fabrication of MFPAs is well advanced even though this technology is only a few years old. Both staring and FLIR line and area arrays have been made and near ideal performance obtained in some cases. Present devices have around 1000 detectors per array, but higher numbers are envisioned in the near future. However, the types of CCD arrays presently being built for visible imagers with detectors in the 10^5 range are not likely to be built soon due to the complexity of the device. Of course, it is not necessary to have that many detectors in FLIR arrays, where optical scanning is used. For example, an array of the type shown in Fig. 4.22b with around 400 columns and 10–20 rows along the TDI direction will give images that are directly compatible with conventional TV displays and the resolution will be comparable to visible imagers. Such an array will have less than 10^4 detectors. For staring arrays, with the advantage of no optical scanning, the number of detectors will have to be appreciably higher for similar resolution.

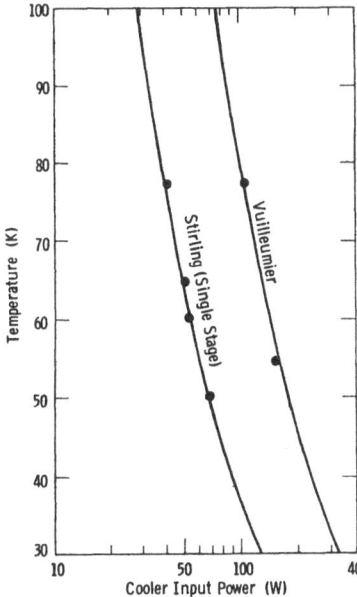

Fig. 4.27. Cooler input power required for a 1 W cold stage heat load as a function of cold stage temperature. The points are experimental data [4.39]

As the size and complexity of the MFPAs increase, the heat dissipation on the focal plane will also increase putting a larger burden on the cooler. Presently, MFPAs have to be operated a few degrees below the temperature of single detectors, which is important from cooler considerations, as shown in Fig. 4.27. Note the rather steep input power increase as the focal plane temperature is reduced. For example, for a 1 W focal plane heat load, typical of advanced extrinsic MFPAs, about 50 W are required for a single stage Stirling cooler at 60 K. If that temperature is reduced to 40 K the power almost doubles. This clearly points out that both cold stage power dissipation and temperature are very important for the cooler not to become excessively power consuming. Further improvements in cooler design are currently underway so that the low operating temperatures of extrinsic MFPAs will not be as much of a disadvantage in the future as they have been in the past.

Acknowledgements. The author is indebted to many of his colleagues at the Westinghouse R&D Center for discussions and some of the results presented here. Special thanks are due to Dr. H. C. Nathanson and Dr. R. N. Thomas. The support of the Institute of Applied Solid State Physics of the Fraunhofer-Gesellschaft in Freiburg, Germany, during the writing of this chapter is gratefully acknowledged.

References

4.1   W. Smith: J. Soc. Tel. Eng. **2**, 31 (1873)
4.2   For a good account of early infrared work, see R. D. Hudson, J. W. Hudson (eds.): *Infrared Detectors* (Dowden Hutchinson and Ross, Stroudsburg 1975) pp. 1–13

4.3   G.E.Stillman, C.M.Wolfe, J.O.Dimmock: "Far-infrared Photoconductivity in High Purity GaAs", in *Semiconductors and Semimetals*, Vol. 12, ed. by R.K.Willardson, A.C.Beer (Academic Press, New York 1977) pp. 169–290

4.4   E.Burstein, J.T.Oberly, J.W.Davisson: Phys. Rev. **82**, 764 (1951)

4.5   B.V.Rollin, E.L.Simmons: Proc. Phys. Soc. (London) B **65**, 995 (1952)

4.6   E.Burstein, J.W.Davisson, E.E.Bell, W.J.Turner, H.G.Lipson: Phys. Rev. **93**, 65 (1954)

4.7   E.E.Godik, Y.E.Pakrovskii, K.I.Svistunova: Sov. Phys. – Semicond. **4**, 624 (1970)

4.8   R.A.Soref: J. Appl. Phys. **38**, 5201 (1967)

4.9   R.W.Redington, P.J. van Heerden: J. Opt. Soc. Am. **49**, 997 (1959)

4.9a A.F.Milton: "Charge Transfer Devices for Infrared Imaging", in *Optical and Infrared Detectors*, ed. by R.J.Keyes, Topics in Applied Physics, Vol. 19 (Springer, Berlin, Heidelberg, New York 1977) pp. 197–228

4.10 A very comprehensive review of extrinsic materials and detectors is P.R.Bratt: "Impurity Germanium and Silicon Infrared Detectors", in *Semiconductors and Semimetals*, Vol. 12, ed. by R.K.Willardson, A.C.Beer (Academic Press, New York 1977) pp. 39–142
      Extrinsic Si is further discussed by N.Sclar: Infrared Phys. **16**, 435 (1976); **17**, 71 (1977)

4.11 M.M.Blouke, E.E.Harp, C.R.Jeffus, R.L.Williams: J. Appl. Phys. **43**, 188 (1972)

4.12 A.F.Milton, M.M.Blouke: Phys. Rev. **3** B, 4312 (1971)

4.13 P.W.Kruse, L.D.McGlauchlin, R.B.McQuistan: *Elements of Infrared Technology*, (Wiley, New York 1969)

4.14 J.S.Blakemore: *Semiconductor Statistics* (Pergamon Press, New York 1962) p. 135

4.15 G.Lucovsky: Solid State Commun. **3**, 299 (1965)
      H.B.Bebb, R.A.Chapman: J. Phys. Chem. Sol. **28**, 2087 (1967)
      V.I.Belyavskii, V.V.Shalimov: Sov. Phys. – Semicond. **11**, 884 (1977)

4.16 R.Baron, M.H.Young, J.K.Neeland, O.J.Marsh: Appl. Phys. Lett. **30**, 594 (1977)

4.17 R.N.Thomas, T.T.Braggins, H.M.Hobgood, W.T.Takei: J. Appl. Phys. **49**, 2811 (1978)

4.18 D.K.Schroder, R.N.Thomas, J.C.Swartz: IEEE Trans. ED-**25**, 254 (1978)

4.19 V.N.Abakumov, V.I.Perel, I.N.Yassievich: Sov. Phys. – Semicond. **12**, 1 (1978) and references cited there

4.20 H.M.Janus, O.Malmros: IEEE Trans. ED-**23**, 797 (1976)

4.21 E.M.Conwell: Proc. IRE **46**, 1281 (1958)

4.22 E.M.Conwell: Proc. IRE **40**, 1327 (1952)

4.23 T.C.McGill, R.Baron: Phys. Rev. B **11**, 5208 (1975)

4.24 C.Jacoboni, C.Canali, G.Ottaviani, A.A.Quaranta: Solid State Electron. **20**, 77 (1977)

4.25 D. F. Barbe: Proc. IEEE **63**, 38 (1975)

4.26 K.Nummedal, J.C.Fraser, S.C.Su, R.Baron, R.M.Finnila: Proc. Int. Conf. Applic. CCDs San Diego (1975) p. 19

4.27 P.Norton: J. Appl. Phys. **47**, 308 (1976);
      P.Norton, R.E.Slusher, M.D.Sturge: Appl. Phys. Lett. **30**, 446 (1977)

4.28 R.D.Nelson: Appl. Phys. Lett **25**, 568 (1974)

4.29 M.R.Hess, J.C.Fraser, K.Nummedal, B.J.Tilley, R.D.Thom: Proc. IRIS **19**, 171 (1974)

4.30 J.T.Longo, D.T.Cheung, A.M.Andrews, C.C.Wang, J.M.Tracy: IEEE Trans. ED-**25**, 213 (1978)

4.31 W.Grant, R.Balcerak, P. van Atta, J.T.Hall: Proc. Int. Conf. Applic. CCDs San Diego (1975) p. 53

4.32 R.W.Swanson, J.Meindl: IEEE Trans. SC-**7**, 146 (1972)

4.33 M.H.White, D.R.Lampe, F.C.Blaha, I.A.Mack: IEEE Trans. SC-**9**, 1 (1974)

4.34 L.M.Candell, M.Y.Pines: Unpublished Hughes Aircraft report

4.35 J.M.Lloyd: *Thermal Imaging Systems* (Plenum Press, New York 1975) Chap. 8

4.36 W.Scott: Appl. Phys. Lett. **32**, 540 (1978)

4.37 R.Baron, J.P.Baukus, S.D.Allen, T.C.McGill, M.H.Young, H.Kimura, H.V.Winston, O.J.Marsh: Appl. Phys. Lett. **34**, 257 (1979)

4.38 T.T.Braggins, H.M.Hobgood, J.C.Swartz, R.N.Thomas: IEEE Trans. ED-**27**, (1980) to be published

4.39 P.LoVecchio, P.Raimondi, R.Longshore: IRIS Det. Spec. Group on IR Det. 145 (1976)

5. Signal Processing with Charge-Coupled Devices[1]

D. F. Barbe, W. D. Baker, and K. L. Davis

With 44 Figures

This chapter deals with the use of charge-coupled devices for performing signal-processing functions. The chapter is divided into two parts: 1) devices and 2) applications. Both analog and digital devices based on the charge-coupled concept are discussed. These device elements include: serial input/serial output, serial input/parallel output and parallel input/serial output building blocks, split-electrode transversal filters, programmable filters, adaptive filters, and digital charge-coupled logic. Also, high-speed CCD technology and CCD/SAW (surface-acoustic wave) device combinations are discussed. Among the applications which are discussed are: electro-optical, sonar, voice communications, and radar. Finally, projections of the application of these devices in future systems are discussed.

5.1 Overview

There is a great deal of interest in the use of charge-coupled devices to perform a wide variety of signal processing functions. The CCD is inherently a sampled analog device, i.e., the number of carriers in a charge packet represents the value of a signal sample. The simplest CCD signal processing device is the serial input, serial output CCD, i.e., the delay line. Charge-transfer devices (the CCD and the bucket brigade device) represent the first analog delay-line devices fabricated in silicon using integrated-circuit processing techniques. Although the CCD delay line is a useful device, the complexity of signal processing functions which can be accomplished with CCDs is much greater than just time delay, as will be described below.

A charge packet in a CCD structure can be made to move in any direction (along the plane of the device) provided that a deeper potential well is induced adjacent to the charge packet in the desired direction. For example, serial-to-

---

1 This chapter is partially based on, and contains, an adapted version of the paper "Signal Processing with Charge-Coupled Devices," D. F. Barbe, W. D. Baker, K. L. Davis, IEEE Trans. ED-**25**, 108–125 (1978) (with permission. Copyright 1978 by the Institute of Electrical and Electronics Engineers, Inc.). The chapter also contains excerpts from the paper "CCD Analog Adaptive Signal Processing," M. H. White, I. A. Mack. G. M. Borsuk, D. R. Lampe, F. J. Kub, 1978 International Conference on the Application of Charge-Coupled Devices Proceedings, San Diego, California, USA, 25–27 October 1978, pp. 3 A-1 to 3 A-14 (with permission. Copyright 1978 by the 1978 International Conference on the Application of Charge-Coupled Devices).

parallel and parallel-to-serial transformations of an array of charge packets are straightforward to accomplish. The parallel input/serial output (PI/SO) device is useful as a multiplexer and also to provide the delay-and-add function.

An important feature of the CCDs is that the transfer of analog samples is controlled by digital clock circuits. Since digital circuits can be designed to produce any desired timing sequence, a great amount of design flexibility is inherent in CCD signal processing technology.

The ability to sense the amount of charge in a sample without disturbing the sample is also important. This is accomplished by using floating electrodes (or diffusions) to couple capacitatively to the charge packets. This signal tapping capability is very useful. For example, if a CCD delay line is tapped at each stage and the taps are weighted, and the weighted samples summed, then the sum is the convolution of the signal with the weighting function, i.e., the device is a transversal filter.

In summary, the sampled analog nature of the CCD coupled with the ability to tap the samples and to control the motion of the samples with digital clocking waveforms make the CCD technology capable of performing a wide variety of signal processing functions.

The use of CCD in a digital mode is also valuable. It is obvious that if the inputs are restricted to nearly full packets (ONEs) and nearly empty packets (ZEROs), the CCD becomes a digital device. It is important to note that signal refresh is possible in a digital device. This removes the limitations on the operation of analog CCDs imposed by leakage current and transfer in-efficiency. Digital arithmetic functions such as adders and multipliers are straightforward in digital CCD structures. The high functional density of analog CCDs is retained in digital CCDs because the cells may be much smaller in digital CCDs as compared with analog CCDs.

The purpose of this chapter is to discuss applications of charge-coupled devices in signal processing systems. The chapter is composed of two parts: 1) a discussion of the various types of CCD signal processing devices and 2) the applications of these devices. In addition to this discussion of analog and digital device structures, sections on high-speed CCDs and CCD-SAW devices are included. In the application part, electrooptical systems, communications, sonar, and radar are discussed. Finally, conclusions and projections are given.

5.2 CCD Signal Processing Devices

5.2.1 Analog Sampled Data Devices

One of the most interesting features of charge-coupled devices is their ability to perform coherent analog signal processing and memory functions. The combination of that ability with low power requirements and high device density makes CCDs attractive for certain systems applications.

The essential features of CCDs for analog signal processing include the input of charge to the CCD, the transfer of charge along the channel, and the conversion of charge to an output signal (current or voltage). The charge transfer process is reasonably well understood (see [5.1] for a summary of current models). With current silicon technology, transfer inefficiencies in the range of 1×10^{-4}–1×10^{-5} can be achieved with clock frequencies up to about 10–20 MHz. Some deep buried channel devices have been clocked at 200 MHz [5.2], with some results suggesting that a clock rate of 900 MHz is possible in silicon [5.3]. A number of electrical input schemes have been developed [5.4–8]. Input methods differ with respect to linearity, noise, speed, and sensitivity to threshold variations. The optimal input technique depends to a large extent on the specific application. Output circuits for CCDs involve the use of floating diffusions or floating gates, which are usually periodically reset to a reference potential, to detect the signal charge [5.9]. In silicon devices, the on-chip circuitry generally includes the detection circuit, a reset switch, and a buffer amplifier with a sample-and-hold circuit sometimes included. The maximum output data rate, dynamic range, and linearity of a CCD signal processing element are often limited by the characteristics of the output circuit.

Serial Input/Serial Output, Parallel Input/Serial Output, and Serial Input/Parallel Output Structures

Basically, CCD signal processing elements can be configured in three ways: with a serial input and output (SI/SO), with a set of parallel inputs along the channel and a serial output (PI/SO), with a serial input and a set of parallel outputs along the channel (SI/PO). These configurations are shown schematically in Fig. 5.1. A particular device may actually incorporate more than one of these structures.

The SI/SO structure of Fig. 5.1a has general applications as a coherent analog delay, as the delay element in recursive filters, and as a variable data rate buffer for time-base compression or expansion. These devices can have charge dynamic ranges on the order of 10^4–10^5 and delay-bandwidth products of a few thousand. Careful attention to the input and output techniques is required in order to realize the large dynamic range with good linearity and to realize high clocking frequencies [5.10, 11]. In addition, the fact that the CCD operates as a sampled data processor may require attention to such problems as aliasing [5.12] and aperture time [5.13] effects.

The PI/SO structure of Fig. 5.1b has applications as a multiplexer, a time-delay-and-integration (or delay-and-add) processor, or for transversal filters. Output problems are basically the same as for the SI/SO devices. Input circuits are similar to those used in SI/SO devices, with the additional complication that threshold voltages (and so offsets) will vary across the inputs. It is possible for the dynamic range of these devices to be limited by the offset pattern noise in multiplexer-type applications.

The SI/PO structure of Fig. 5.1c is primarily applicable to the realization of transversal filters. It has been used for spectral filtering, transform calculation,

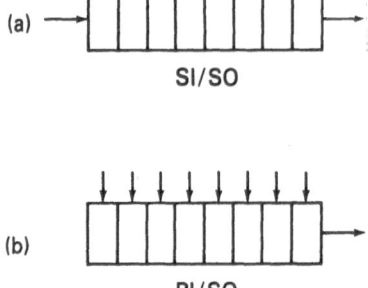

(a)

SI/SO

(b)

PI/SO

(c)

SI/PO

◀ **Fig. 5.1a–c.** Basic CCD signal processing configurations: (a) serial input and output, (b) parallel input with serial output, and (c) serial input with parallel output

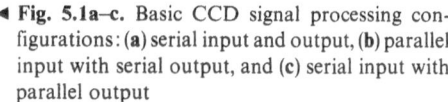

INPUT

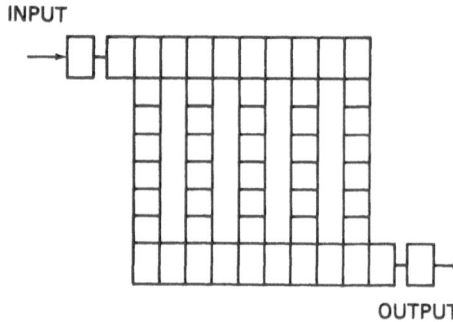

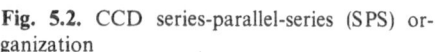

OUTPUT

Fig. 5.2. CCD series-parallel-series (SPS) organization

and convolution/correlation. Critical issues for this device are the nondestructive tapping of the signal and the appropriate techniques for weighting the tapped signal. Dynamic range and/or operating speed may well be limited by the tapping and weighting circuitry.

A combination of CCD basic structures can easily be incorporated into a single device. Figure 5.2 illustrates a series-parallel-series (or SPS) device which could be used in a frame store or similar application. This device involves SI/PO, SI/SO, and PI/SO structures, but once past the initial input circuit the signal is handled entirely as charge packets, and interface circuits between the different portions of the device are not necessary unless dc offset correction is required at some point.

The Split-Electrode Transversal Filter
The split-electrode transversal filter is such a versatile signal processing element that it deserves special attention. The split electrode technique [5.14] provides an elegant way to implement the multiply-and-add functions required in a transversal filter. The principle of operation of the split electrode technique is that as charge transfers within the silicon into the region under an electrode, an opposite charge must flow onto the electrode from the clock line. On one phase, the electrodes are all split into two sections of varying sizes as shown in Fig. 5.3. Figure 5.3 shows the ϕ_3 electrode being split, but any of the phases may be used. One side of each split electrode is connected to the ϕ_3^+ clock line and the other to the ϕ_3^- clock line. The two ϕ_3 lines are clocked simultaneously with the same clock phase. The signal-dependent charge which flows into each portion of a split electrode is proportional to the area of that part of the electrode and to the signal charge flowing into the region under that electrode. By measuring the difference in charge in the two sections of a split electrode, the nonde-

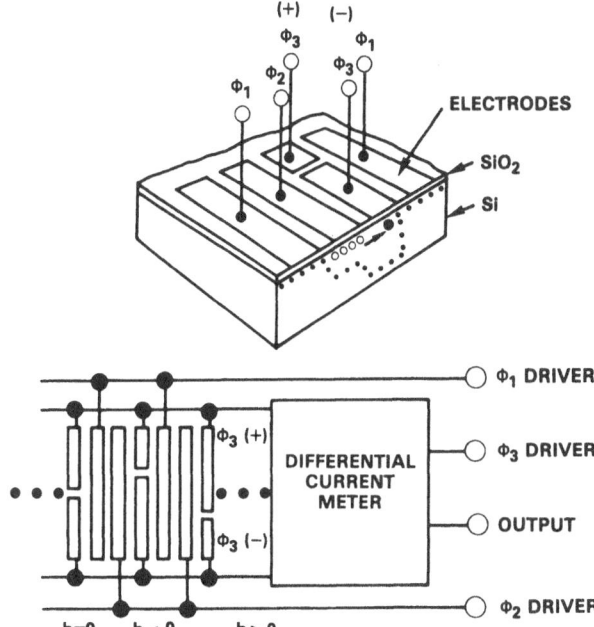

Fig. 5.3.
Schematic diagrams of analog CCD split-electrode
transversal filter

structive sampling and weighting operations are performed. Since the ϕ_3^+ electrodes are tied together and the ϕ_3^- electrodes are tied together, the summation occurs automatically, and the output signal at each clock period is proportional to the difference in current flowing in the two lines of split phase. If the split in the mth electrode occurs exactly in the center of the CCD channel, the tap weight is $h_m = 0$. If the electrode is split at a position such that all of the charge flows into the ϕ_3^+ or ϕ_3^- clock lines, the resulting weighting coefficient is $h_m = +1$ or $h_m = -1$. Intermediate values are possible, the resolution being limited by tolerances of masks and lithography [5.14a]. The filter output is obtained by intergrating the difference current which flows in the ϕ_3^+ and ϕ_3^- clock lines. This is performed in a differential current integrator (DCI) as shown in Fig. 5.3.

The performance of split electrode transversal filters is limited by imperfect charge transfer, weighting coefficient error, noise, and nonlinearity. The effect of imperfect charge transfer is a shift in the frequency response of the filter [5.15]. In a split electrode filter the weighting coefficients are coded into the metal mask as gaps in the gate electrode structure, and the predominant weighting coefficient error then becomes the location of these gaps on the photomasks. When computer generation is used these gaps are located at quantized intervals δ. Therefore, the fractional quantization step δ/W is dependent on the channel width W. The ratio of the error response Δ_{rms} to the correlated signal response v_0 is on the order of $\Delta_{rms}/V_0 = (\delta/W)/\sqrt{(3N)}$, where N is the number of stages in the filter. For practical CCD filters this ratio typically ranges between 60–80 dB. CCDs are very low-noise devices, and in a CCD transversal filter, this noise is effectively reduced even further by the processing gain of the filter. However, because of the effective insertion loss which results before amplification, the differential amplifier and associated output circuitry in the DCI must have very low noise if the ultimate filter dynamic range is to be achieved. A dynamic range of 75–80 dB (defined as the maximum peak-to-peak signal with acceptable distortion to *rms* noise level) has been achieved using 500-stage filters [5.16]; by suitably designing the output amplifier, 100 dB is predicted. Very high linearity is difficult to achieve with CCD transversal filters. However, total harmonic distortion of less than 1 percent has been achieved on a 500-stage filter for input voltage of up to 5 V $p-p$ [5.16]. The nonlinearities are due to imperfect sampling of the input signal voltage and the nonlinear depletion layer capacitance.

The CCD split-electrode transversal filter is especially useful in performing the Fourier transform using the chirp-z transform (CZT) algorithm. The CZT gets its name from the fact that it can be implemented in an analog manner by 1) premultiplying the time signal with a chirp (linear FM) waveform, 2) filtering with a chirp convolution filter, and 3) post-multiplying with a chirp waveform. When implemented digitally, the CZT has no advantages over the conventional fast Fourier transform algorithm [5.17]. However, the algorithm lends itself naturally to implementation with CCD split-electrode transversal filters [5.18]. Starting with the definition of the DFT

$$F_k = \sum_{n=0}^{N-1} f_n e^{-i2\pi nk/N} \tag{5.1}$$

and using the substitution

$$2nk = n^2 + k^2 - (n-k)^2 \tag{5.2}$$

the following equation results:

$$F_k = e^{-i\pi k^2/N} \left[\sum_{n=0}^{N-1} (f_n e^{i\pi n^2/N}) e^{i\pi(k-n)^2/N} \right]. \tag{5.3}$$

This equation has been factored to emphasize the three operations which make up the CZT algorithm, i.e., premultiplication, filtering, and post-multiplication. When only the power density spectrum is required, the post-multiplication by $\exp(-i\pi k^2/N)$ can be eliminated; a block diagram of the circuit implementation is given in Fig. 5.4. This block diagram has been implemented using 500-stage CCD filters to obtain the sliding DFT [5.16]. Using this device on a 200 Hz square-wave input, up to 49 harmonics were observed before the noise level obscured the output.

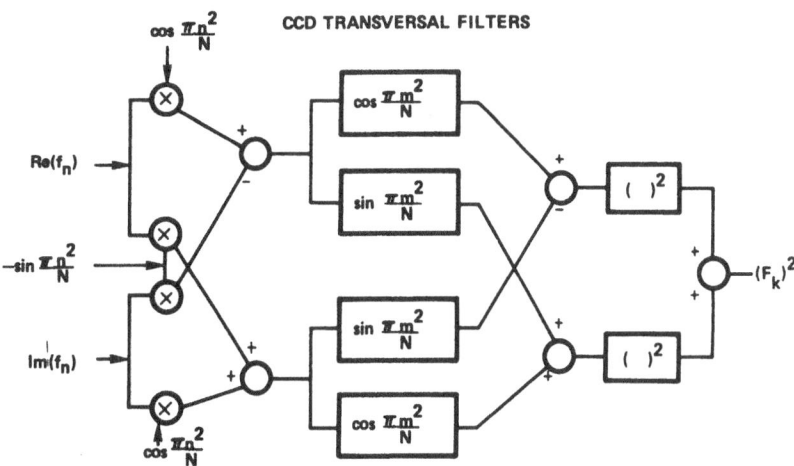

Fig. 5.4. Block diagram of the complex arithmetic of the chirp-z transform (CZT) algorithm for computing power spectral density spectrum. The transversal filters are designated by COS and SIN

The Double-Split-Electrode Transversal Filter [5.19]

A new approach to the split electrode transversal filter employs two splits in the CCD sensing electrodes to provide better performance than the conventional single-split weighting technique.

The main disadvantage of the conventional split-electrode weighting technique is the fact that the entire area of each electrode is used in determining the value of each coefficient. In the case of small taps, the gaps are located near the middle of the electrode so that a small tap value is obtained from the difference of two relatively large values. Only a very small portion of the area of the electrode contributes to the actual tap weight while a much larger area contributes to an undesirable common mode signal. Since most transversal filters have a large proportion of small tap weights, the common mode components accumulate to produce a signal which can be an order of magnitude higher than the actual output signal. This excess common mode signal is just one aspect of the problems associated with excess capacitance on the sensing busses of the filter. The double-split-electrode structure has been applied to reduce the problems associated with excess capacitance at the sensing nodes.

In the double-split-electrode structure, there are two splits per sensing electrode, partitioning it into three segments. The two outer segments are used to define the positive and negative coefficients. The middle segment is clocked separately from the outer segments and represents excess capacitance, not required in the determination of the tap values. Removal of this excess capacitance has four important consequences to the filter operation: 1) suppression of common mode signal, 2) reduced sensitivity to gain error in the sensing amplifiers, 3) reduced noise gain in the sensing amplifier, and 4) reduced noise pickup from the sensing electrodes.

The following example will serve to illustrate the advantages associated with reducing the capacitance on the filter sensing nodes. The basis of this example is a 32-tap low-pass filter. Most of the taps are much smaller than the maximum coefficient such that when this filter is implemented using the conventional split-electrode approach only 16 % of the sensing electrode area contributes to the output signal. The other 84 % of the electrode area represents unwanted capacitance which is connected to the sensing nodes. By using the double-split-electrode structure this unwanted capacitance can be removed from the sensing nodes. For the conventional single-split-electrode case, the common mode signal at dc is almost 8 times the magnitude of the filter output signal. This can place severe constraints on the maximum signal level since the inputs to the sensing amplifiers have to handle the full excursion of the positive and negative sensing outputs. For the double-split-electrode case, the common mode signal at dc is approximately equal to the signal magnitude, thus putting much less constraint on the sensing amplifier. The second advantage which was listed was reduced sensitivity of the output to gain error in the individual sensing lines. For the low-pass filter case the double-split-electrode structure is less sensitive to gain error by a factor of 5. In typical configurations the gain of the sensing amplifiers is proportional to the input capacitance (C_i). This implies that noise gain from the sensing electrodes will be C_i/C_{fb} and the noise gain of the amplifier will be $1 + C_i/C_{fb}$, where C_{fb} is the feedback capacitance. Reducing of C_i (by a factor of 5 in this example) can have a significant effect on the output noise level and hence the available dynamic range of the filter. One final advantage listed for the smaller sensing capacitance was reduced noise pickup from the clocks and other sources on chip. There are other advantages to the new structure which can be attributed to the presence of two splits instead of one. The error due to tap quantization can be considered for the two cases. For the single-split approach an error in position, Δx, will result in Δx increase in the length of one side and Δx decrease in the length of the other side. The output error is then proportional to $\Delta x - (-\Delta x) = 2\Delta x$. For the double-split approach, the two positions can be considered as independent and the expected output error will be proportional to $\sqrt{2}\Delta x$. Similar reasoning shows that an improvement of $\sqrt{2}$ is made for partitioning noise which occurs when the charge must be split into the different segments of the sensing electrodes.

A 71-tap low-pass filter using the double-split-electrode approach has been implemented. There are practical considerations which affect the structure. For example, the large taps do not have sufficient space to put two splits in the electrode and are implemented as single-split taps. Two parallel channels have been added to equalize the capacitance on both sensing busses and assure complete symmetry of the CCD. Also the sensing segments have a finite minimum size to ensure that edge effects on both sides of the channel are minimized and symmetrical. The frequency response of this filter has been measured and found to agree with the simulated results with less than 0.1 dB ripple in the passband and better than 40 dB rejection in the stopband. The dynamic range of the filters was measured to be better than 83 dB.

Programmable Transversal Filter

When a fixed filter function is required, a CCD split-electrode transversal filter offers a low-power, high-density, and, if quantities are large enough, a low-cost approach. If, however, quantities of the specific filter are not sufficiently large to support a specialized mask or if the application requires a programmable filter function, the split-electrode filter is not suitable.

Such applications can, in principle, be satisfied with a CCD analog correlator consisting of two tapped CCD shift registers with analog multipliers connected to each pair of taps. In practice, such correlators have not been useful primarily because of problems with the analog multipliers. Nonuniform offsets and gains of the multipliers and power dissipation of the multipliers have been the primary problems. Also, the on-chip storage time of the analog reference channel, in this approach, is limited.

The preferred approach to programmable CCD transversal filters is to use a CCD in the configuration shown in Fig. 5.5. This scheme uses a CCD to perform a serial-to-parallel transformation, to provide N delay channels, and to provide a summing bus (diffusion) for the outputs of the N channels. A sample-and-hold circuit and a multiplying digital-to-analog converter (MDAC) are also needed [5.20]. The weighting coefficients can be obtained in digital form from read-only memory or as calculated by a microprocessor (for an adaptive system).

If τ is the delay per stage in the delay columns and $f_c \equiv 1/\tau$, then the sample-and-hold period is τ, the sequence of weighting coefficients $W_1, W_2, ..., W_N$ are read into the MDAC in time τ, and the products $W_1 V_i(t), W_2 V_i(t), ..., W_N V_i(t)$ are read into the serial CCD register in time τ. Then, this set of analog samples is shifted down the delay columns as new sample sets follow. As the delayed samples reach the ends of the columns, they are clocked onto a diffused diode extending across the N columns when the samples are summed. The output is then

$$V_0(t) = \sum_{n=1}^{N} W_n V_i(t - n\tau), \tag{5.4}$$

which is the transversal filter equation.

Thus, this approach is an alternate to the split-electrode transversal filter, and it is particularly useful when programmable weighting coefficients are desired. The main disadvantage is the high speed at which the MDAC has to operate; i.e., the MDAC must operate at a speed which is N times the sampling rate. Therefore, this approach is presently limited to low-bandwidth applications.

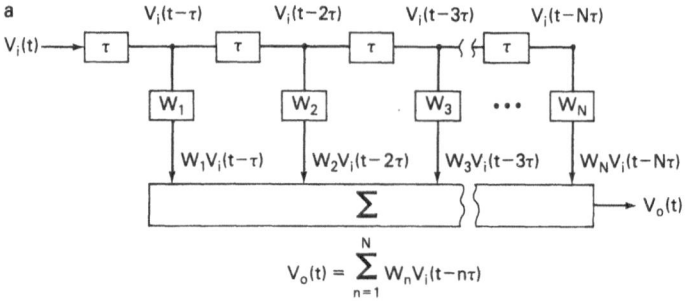

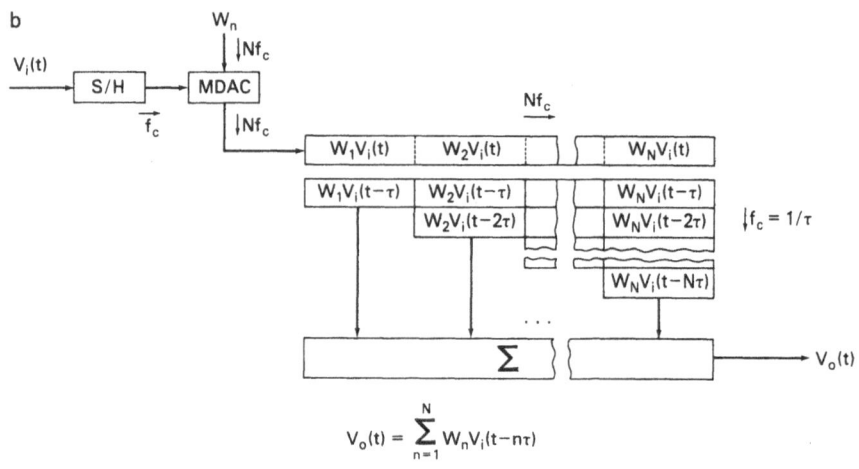

Fig. 5.5a, b. Schematic diagrams illustrating implementations of a transversal filter using (**a**) a tapped and weighted CCD delay line and (**b**) a parallel input CCD array

The Adaptive Transversal Filter [5.21]

Figure 5.6 is a block diagram of an adaptive filter. For sampled-data input signals, an error is formed at each clock according to the expression

$$\varepsilon(m) = d(m) - y(m), \tag{5.5}$$

where d is the desired input, m is the clock index, and y is the weighted sum of the past N inputs, with N the length of the delay line. The error is used as the input to the algorithm which in turn adjusts the weight at each tap location to minimize the mean-square error.

The "clipped data" algorithm changes the weight of each tap location according to the equation

$$W_i(m+1) = W_i(m) + 2\mu\varepsilon(m)\mathrm{sgn}X(m)$$

$$= W_i(0) + 2\mu \sum_{k=1}^{m} \varepsilon(k)\mathrm{sgn}X_i(k), \tag{5.6}$$

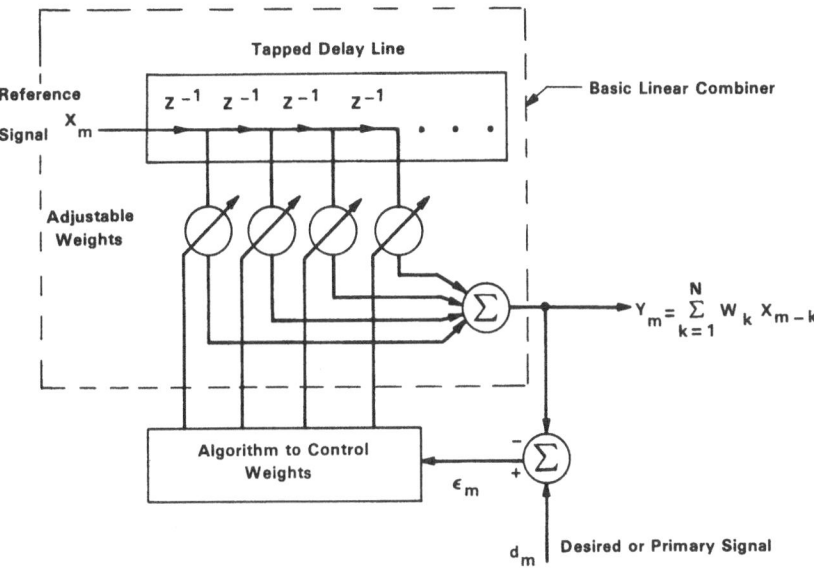

Fig. 5.6. Schematic diagram of an adaptive transversal filter [5.21]

where i is the tap location, μ is a constant which determines stability and convergence rate, ε is the instantaneous error defined by (5.5), X is the tap output, and sgn is the sign function. Equation (5.6) indicates that the algorithm retains full linearity of the error but requires a multiplier and integrator at each tap location. The multiplier is actually a branch operation which checks the sign of X and on this basis adds or subtracts the quantity $2\mu\varepsilon(m)$ from the current tap weight to form the new weight value.

A block diagram of the monolithic analog adaptive filter is shown in Fig. [5.7]. The analog delay line is a CCD tapped delay line with a floating clock electrode sensor circuit at each tap location to nondestructively sense the CCD signal charge and provide charge-to-voltage conversion [5.22]. The output of each tap location is a voltage which is converted to a weighted current, representing the required multiplication, via the drain-source conductance of an NMOS transistor biased in the triode region. In the triode region, the incremental drain-source current is given by

$$i_{ds} = K(V_{GS} - V_T)V_{ds}(V_{ds} \cong 0)$$
$$= g_{ds}V_{ds}, \qquad (5.7)$$

where K is a constant dependent on device geometry and processing parameters. V_{GS} is the gate-source voltage (integrated value of $2\mu|\varepsilon|$), V_T is the threshold voltage, V_{ds} is the drain-source voltage (delay line output), and g_{ds} is the drain-source conductance. Positive and negative weight values are achieved by using *two transistors* at each tap location, setting the conductance of one transistor to

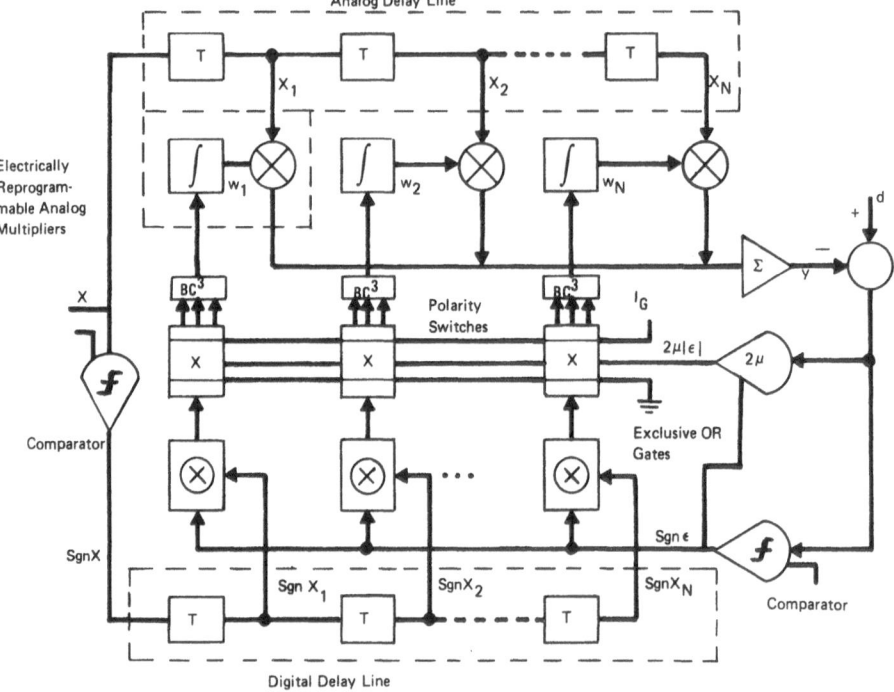

Fig. 5.7. Schematic diagram of a monolithic CCD analog adaptive filter [5.21]

a fixed value, and allowing the conductance of the second device to vary via the adaptive algorithm. The required integrator is realized via the gate capacitance of the NMOS transistor weight. The incremental change in tap weights is achieved using the bidirectional charge controlled circuit described below.

The two currents from each tap are summed simultaneously with the weighted currents from the respective transistors at the other tap locations. The input nodes of two CMOS operational amplifiers serve as the two summing nodes. The two currents are converted to a voltage and further processed by a subtractor, comparator, absolute value circuit, and gain amplifier, as required by the "clipped data" LMS algorithm.

Refering to (5.6), the weight update algorithm requires a comparator at each tap location to form sgn $\{X(m)\}$. The circuit complexity is reduced considerably by using a single comparator at the delay-line input and applying the comparator output (sgn X) to a digital shift register which is clocked in synchronism with the analog sample in the CCD. An exclusive OR circuit provides the digital multiplication which provides the branch operation to increment or decrement the voltage of the NMOS weight.

On-chip timing is generated using three D-type flip-flops configured as a ripple counter to provide decoding waveforms for the NAND/NOR gates used in the combinatorial logic. The ability to update all the weights simultaneously relaxes the requirements of the clock drivers.

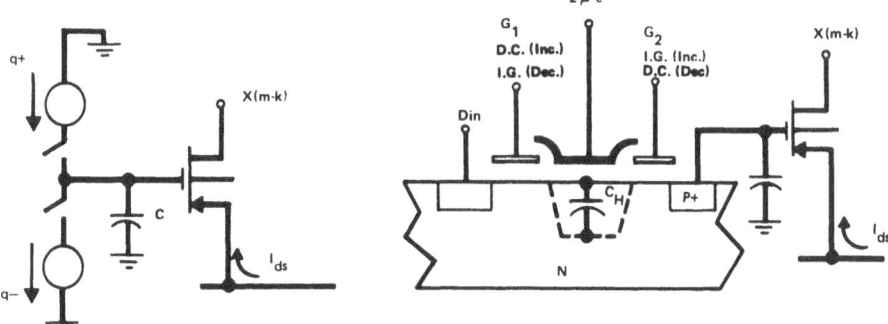

Fig. 5.8. Bidirectional charge control model for incremental adjustment of MOSFET analog conductance weight. To increment the weight, G_1 is operated at DC and G_2 is operated as the input gate. To decrement the weight, G_1 is operated as the input gate and G_2 is operated at DC [5.21]

In order to achieve a variable V_{GS}, the error $\varepsilon(mT)$ can be converted to digital form with an A/D converter, storage and accuracy can be achieved with accumulators, and finally, the multiplication can be accomplished with multiplying digital-to-analog converters (MDACs) [5.23]. This is an attractive approach because of the recent advancements in capacitor-weighted MDACs. However, unless the MDAC is shared over many taps, the complexity of the adaptive signal processor is increased [5.24]. In general, an MDAC for each tap weight is not too appealing an approach because of the chip area involved in the layout. In addition, the off-chip peripheral hardware is quite involved because of storage and accuracy requirements in the A/D conversion process. An 8-tap weight system has been built with this approach and it performed adaptive linear prediction with the clipped LMS algorithm [5.24]. However, the approach used lumped LC delay lines with different dispersive characteristics, and a large amount of peripheral hardware was required for implementation.

Another method for achieving programmable weights is to replace the NMOS transistors described above with metal-nitride-oxide-silicon (MNOS) nonvolatile memory transistors. Memory is achieved in the MNOS transistor by electrically reversible tunneling of charge from the silicon semiconductor to deep traps near the SiO_2/Si_3N_4 interface. By the application of suitable voltages to the gate of the transistor, the threshold voltage can be changed in discrete increments. These changes can then be sensed as a change in the conductance of the transistor. This type of device has been used as the programmable tap weight and integrator, with a trapped CCD, to demonstrate in a hybrid from a 2-tap integrated circuit LMS adaptive filter.

A final approach to achieving programmable tap weights is the bidirectional charge-controlled circuit (BC^3). The technique uses stabilized charge injection to increment or decrement analog signal charge onto the node of a MOSFET analog conductance. The concept is illustrated in Fig. [5.8]. The

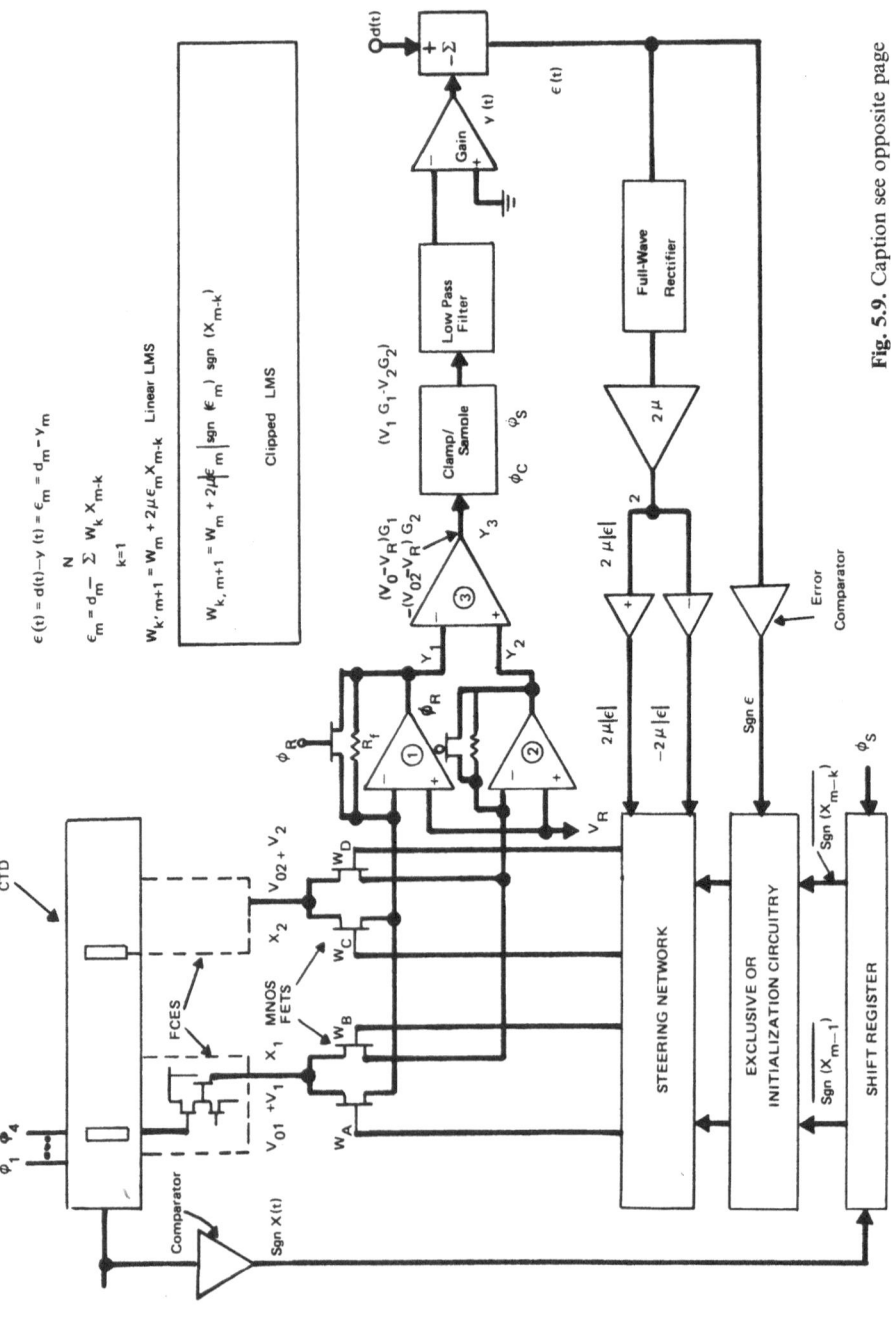

Fig. 5.9. Caption see opposite page

analog, scaled error signal $2\mu\varepsilon$ is applied to a CCD storage or holding well and the control of the signal is accomplished by selecting the proper function of the G_1 and G_2 electrodes. The selection process involves the binary multiplication of

$$\text{sgn}\{\varepsilon(mT)\} x \, \text{sgn}\{x(m-k)\}, \tag{5.8}$$

with an exclusive OR gate as indicate in Fig. [5.7]. The incremental voltage applied to the gate of the MOSFET voltage-controlled analog conductance weight is

$$\Delta V = 2\mu|\varepsilon|\frac{C_{\text{H}}}{C}, \tag{5.9}$$

where $|\varepsilon|$ is formed with an absolute-value amplifier, and C_{H} is the holding well capacitance. The storage or integration of the weight values is performed with the on-chip capacitance C associated with the gate node of the MOSFET weight. In order to achieve long-term weight retention, as might be necessary for some voice processing applications, either the C must be increased, which decreases the sensitivity of the weight to adjustment and increases the chip area for on-chip weighting, or some method of weight updating must be employed. One such method is to use an off-chip weight capacitance. Another method is to employ an A/D converter, memory, and a MDAC to sequentially update the weights in synchronization with the CCD clock.

The feasibility of implementing the "clipped" LMS algorithm has been confirmed by fabricating and testing a 2-tap weight hybrid processor. The processor has been constructed using a basic linear combiner composed of a CCD serial in/serial out structure, MNOS analog conductances, operational amplifiers, comparators, CMOS switches, and CMOS logic. A block diagram of the hybrid processor is shown in Fig. [5.9]. The processor was configured, as shown in Fig. 5.10, to allow characteristic measurements to be made. Measurements were performed to determine the convergence factor μ as a function of processor transient response. The ability of the processor to track as a function of phase and amplitude variation in the desired channel [$d(t)$ of Fig. 5.10] was also confirmed.

The processor was then configured as a noise canceller. The block diagram of the arrangement and experimental results are presented in Fig. 5.11. The desired signal was corrupted by a narrow band tone 16 dB greater. After processing by the adaptive filter, the output interfering tone was 18 dB below the desired tone, representing a rejection of 34 dB.

◄ **Fig. 5.9.** Block diagram of an adaptive signal processor using the clipped data LMS algorithm [5.21]

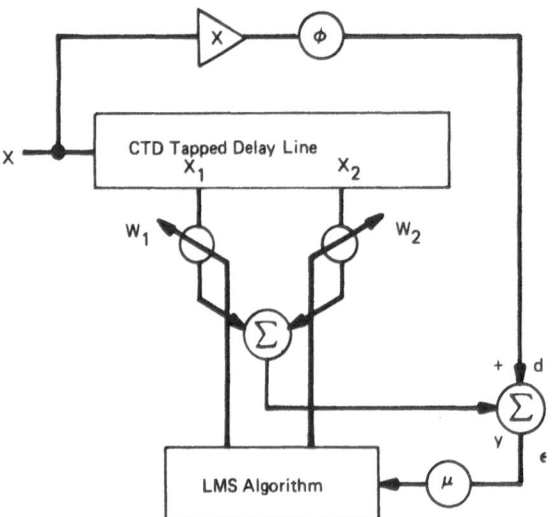

Fig. 5.10. Configuration of a hyprid adaptive processor for making fundamental performance measurements [5.21]

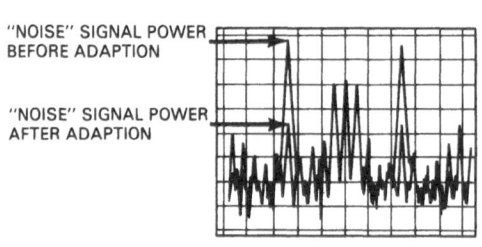

Fig. 5.11. Trace of oscilloscope and block diagram illustrating the CCD adaptive noise canceller with a 2-tap adaptive element [5.21]

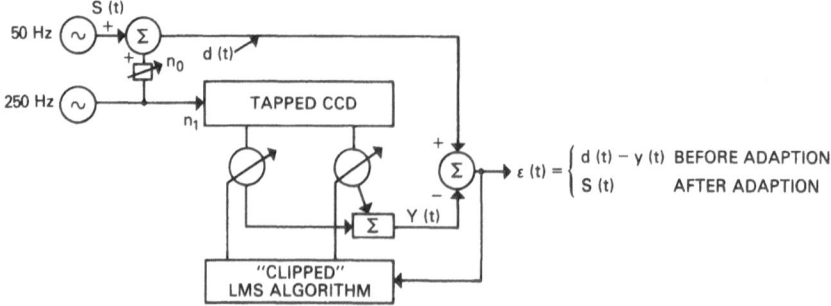

5.2.2 Digital Charge-Coupled Logic (DCCL)

Most of the CCD chips developed up to now for signal processing applications make use of the *analog* sampled data nature of the charge-coupled concept. These devices are quite important and are having a large impact on signal processing techniques. However, there are many signal processing tasks which require digital techniques for various reasons, e.g., more accuracy is required

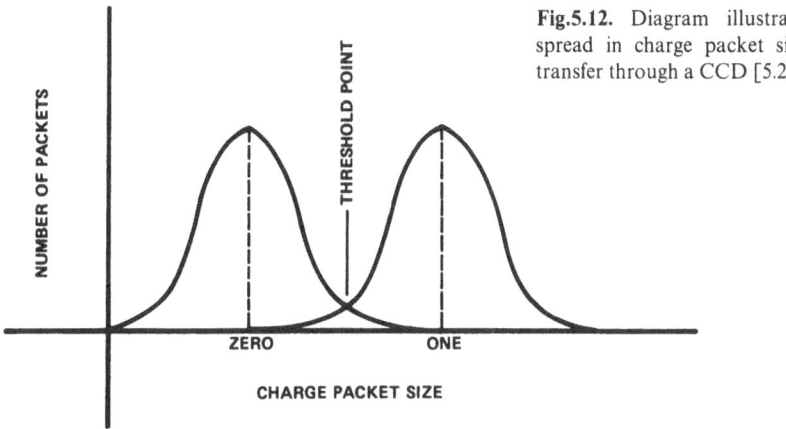

Fig.5.12. Diagram illustrating the spread in charge packet sizes after transfer through a CCD [5.26, 27]

than can be achieved with analog techniques. A digital CCD signal processing technology is being investigated which combines the high-accuracy capability of digital systems with the low-power high-functional density of CCD to form a unique LSI technology [5.25]. Some complex digital systems that were previously thought impractical are now feasible.

Digital memory was one of the first uses suggested for CCDs. This application uses the fundamental transfer characteristic of the device to best advantage. The shift register storage wells are either nearly full (to represent a binary "one") or nearly empty (to represent a binary "zero"). The input circuit, therefore, has to produce only two well-defined states in the shift register. Consequently, input linearity is of no consequence; in fact the input circuit assumes the highly nonlinear characteristic of a threshold circuit. Charge packets arriving at the shift register output will have a continuum of values centered around the two binary states. The lack of precise values for the two states at the output is due to noise in the shift register.

The general case will appear as shown in Fig. 5.12. The density functions for the "one" and "zero" charge packets will have the characteristics of Gaussian distributions, with variances determined by the various noise sources present. The function of the output circuit attached to the shift register is to perform a thresholding operation on these charge packets and to produce a signal capable of initiating in another digital device a clean "one" (or full well) for every charge packet whose amplitude exceeds a given minimum value. Conversely, the circuit should be capable of producing a well-defined "zero" signal for all charge packets whose amplitude is less than a given maximum. Since the output charge packet size can assume any of a continuum of values, the output circuit will inevitably misinterpret some signals; the missed signals will be exactly those in the tails of the distributions which lie on the wrong side of the circuit threshold. These mistakes represent the bit error rate (BER) of the device. Once a charge packet has passed through such a threshold circuit, the information is capable of being reinserted into a shift register and so the charge packet is said

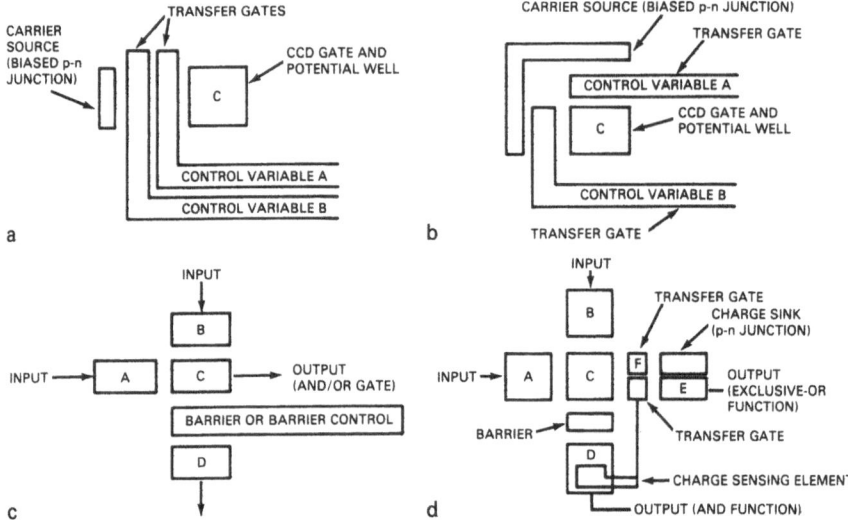

Fig. 5.13a–d. Schematic diagrams illustrating the logic functions which can be derived using transfer gates, carrier sources, and potential wells: (**a**) AND gate, (**b**) OR gate, (**c**) exclusive-OR gate and (**d**) exclusive-OR and AND functions realized simultaneously [5.26, 27]

to be refreshed. This operation can be carried on indefinitely until such time that the accumulated errors make the data no longer useful. The number and placement of such refresh circuits is an important design consideration in any digital CCD processor, for it is these circuits that permit a large number of calculations to be performed while maintaining an acceptable overall BER.

Performing digital-logic functions with the charge-transfer principle requires interaction with the information contained in charge-coupled shift registers. Such interaction can be accomplished in two ways. One is termed bit destructive because in the process the original bits lose their individual identities. The other, designated bit preserving, detects the presence or absence of charge without disturbing the bit stream. This detection controls charge flow in another register.

Generally speaking, any of the schemes used to provide weighted tap points in analog charge-coupled filters can be adapted for use as a nondestructive charge-sensing operation, suitable for bit-preserving logic circuits. In such digital applications, tapping and weighting schemes are relatively easy to implement since the tap values need only be the equivalent of a one or zero.

Although both methods find many applications, there is a significant difference. The bit-preserving method allows repeated operations on the original data stream, since it is always preserved. But the bit-destructive method can operate only once on any bit stream.

In the basic AND gate implemented with CCD logic in Fig. 5.13a, the two shift registers A and B are connected to two series-transfer gates. Both gates must be on for mobile minority carriers to reach the gate marked C. In logic symbols,

$$C = A \cdot B = AND. \tag{5.10}$$

An OR function can be similarly implemented (Fig. 5.13b). The controlled transfer gates are in parallel. With the same reasoning, the equation is

$$C = A + B = OR. \tag{5.11}$$

Both schemes are bit preserving and do not disturb the controlling bit patterns. Bits are simply sensed nondestructively and used as controllers for other registers.

However, there are some useful bit-destructive logic circuits. Providing OR and exclusive-OR gates requires a configuration only slightly more complex. In Fig. 5.13c, shift registers A and B dump their charge packets directly into the potential well under gate C. Another gate, D, is biased in a mode capable of accepting charge, but it is separated from gate C by a potential barrier that is created by ion implantation or controlled by a separate gate voltage.

In either case, the capacity of the potential wells under gates A, B, D, and D are all equal; therefore, if A and B are filled to capacity, their combined charge packets are more than can be contained under C. The barrier then allows this surplus of charge to flow under D. If only A or B is full of charge, C will be filled exactly and D will remain empty. So C represents on OR function and D and AND function,

$$C = A + B = OR$$
$$D = A \cdot B = AND.$$

Carrying the implementation one step further produces an exclusive-OR gate. Figure 5.13d shows that gate D contains a charge-sensing element and drives one of two parallel-transfer gates that control the flow of charge from gate C. The charge-sensing element will allow the charge under gate C to flow to gate E only if there is no charge under gate D. The purpose of transfer gate F is to clear out charge under C if it has not been moved under E. Not only does E provide and exclusive-OR function and D and AND function, but the two gates together provide a complete two-bit half adder with a sum and a carry.

Another basic logic element of digital arithmetic is the full adder [5.26, 27]. A digital CCD implementation of a full adder is shown in Fig. 5.14 which shows three inputs A, G, and B. The barrier heights are chosen to block a single full well of charge. (The barriers are achieved by ion implantation or by the application of a voltage on an electrode.) The charge-sensing device (e.g., floating gate) and the transfer gate T are implemented in such a way that if there is zero charge at location C, then the transfer gate T is turned on, and if there is a full packet of charge at location C, then the transfer gate T does not

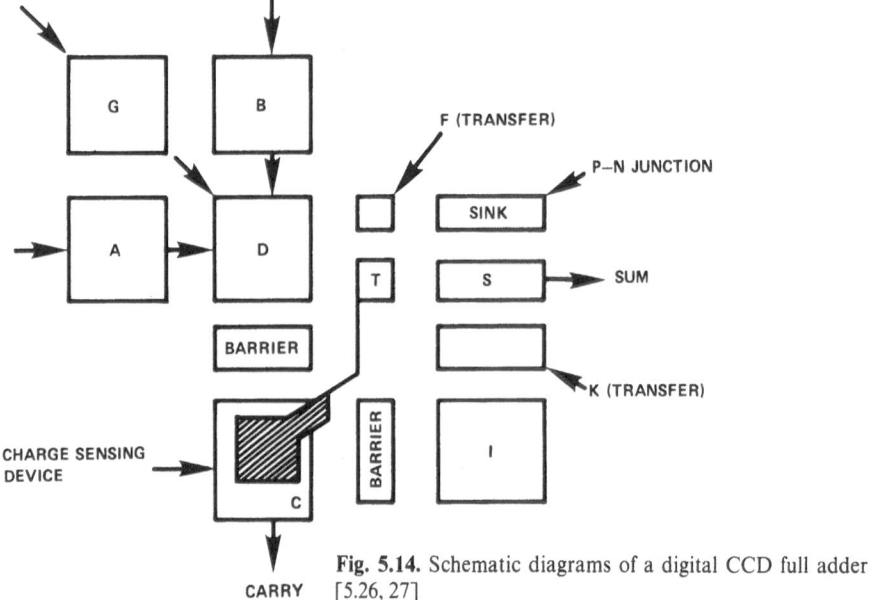

Fig. 5.14. Schematic diagrams of a digital CCD full adder [5.26, 27]

transfer charge from D to S. Transfer gate F is activated after the addition is complete to clear the adder for the next addition.

The circuit operation is based on the fact that gate C will have charge under it only if at least two of three inputs, A, B, G, contain charge. The gate marked I will receive charge only if all three inputs contain charge. The truth table for the various gate locations of Fig. 5.14 is given in Fig. 5.15. Note that the gates S and C do indeed represent the sum and carry functions of the desired full adder. The basic aspects of the timing sequence are shown in Fig. 5.16, where the basic time unit is designated as $1/(2f_0)$ and is intended to represent the time required for sufficient charge to transfer from one storage well to the next. Accordingly, gate I is indicated as being in the storage mode for three time units in order to ensure that the input charge has sufficient time to traverse the necessary distances. Note that the overall time required for a complete operation with the timing diagram shown is $5/(2f_0)$. This means that the full adder will operate at $(2/5)$ of the frequency of the basic shift register. Figure 5.17 shows typical output waveforms achieved by an early full-adder circuit.

The existence of a viable full-adder circuit allows the construction of other more complex digital functions. Consider Fig. 5.18 which diagrams the use of full adders to produce a 16-bit digital adder. Two 16-bit words, represented by the bits $(a_1, a_2, ..., a_{16})$ and $(b_1, b_2, ..., b_{16})$ are simultaneously applied to the inputs shown, and the clocking sequence begins. Appropriate delays are inserted before each full-adder circuit so that input bits arrive at the proper full adder at the same time as the corresponding carry bit arrives from the preceding partial addition. The delays inserted after each full adder are used to

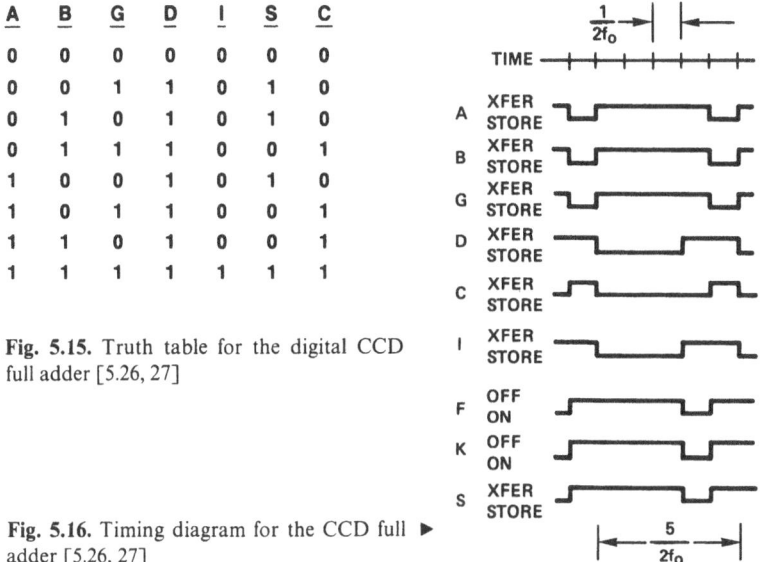

A	B	G	D	I	S	C
0	0	0	0	0	0	0
0	0	1	1	0	1	0
0	1	0	1	0	1	0
0	1	1	1	0	0	1
1	0	0	1	0	1	0
1	0	1	1	0	0	1
1	1	0	1	0	0	1
1	1	1	1	1	1	1

Fig. 5.15. Truth table for the digital CCD full adder [5.26, 27]

Fig. 5.16. Timing diagram for the CCD full ▶ adder [5.26, 27]

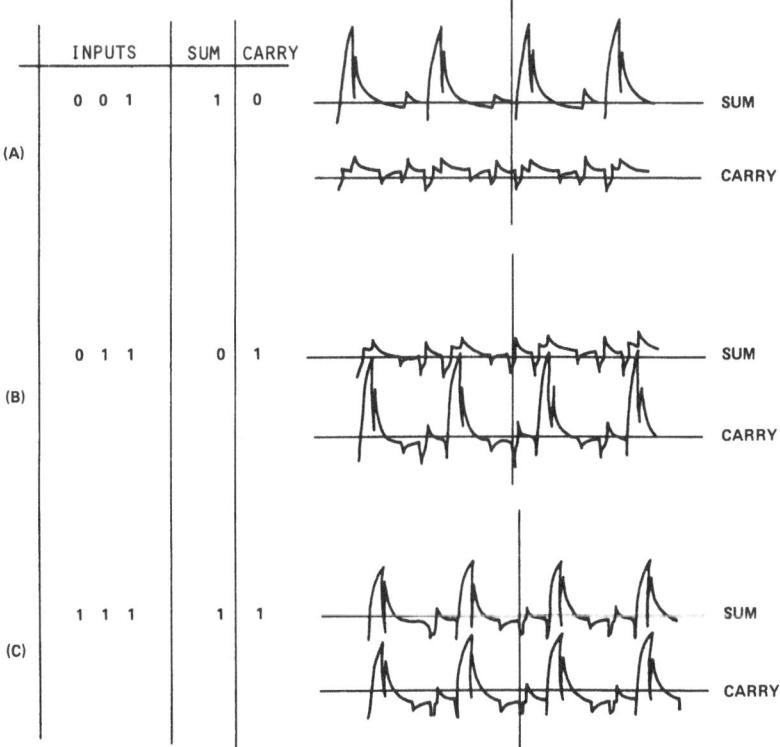

Fig. 5.17. Typical output waveforms from an early CCD full-adder circuit [5.26, 27]

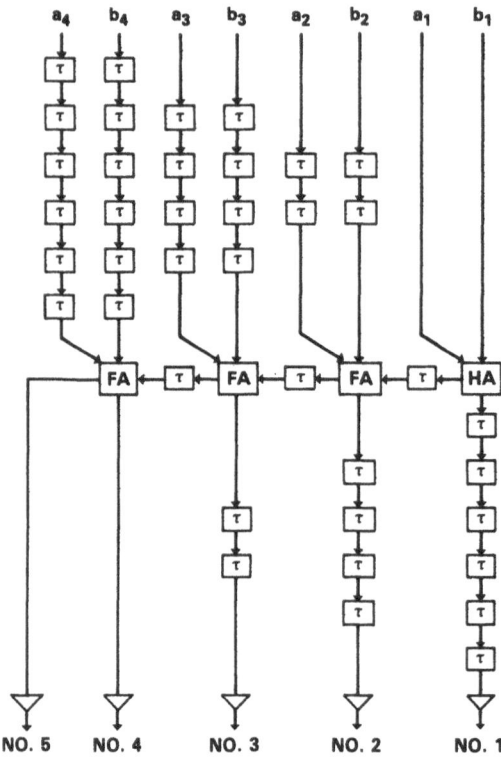

Fig. 5.18. Schematic diagram showing the use of the basic full-adder circuits and CCD delays to produce a two-word, N-bit adder. A four-bit ($N = 4$) adder is illustrated [5.26, 27]

CARRY BITS	c4	c3	c2	
A-WORD	a4	a3	a2	a1
B-WORD	b4	b3	b2	b1
SUM	c5	S4	S3	S2	S1

store intermediate results until all calculations are completed and each bit of the entire sum word arrives simultaneously. This procedure means that there will be an initial throughput delay when new data are entered into the 16-bit adder, but it also guarantees that new inputs can be applied with minimum spacing in time so that after the initial delay, output results are available on a "continuous" basis, with the delay between results being only $5/(2f_0)$ seconds. A similar organization can produce a digital multiplier. Figure 5.19 diagrams the arrangement of an expandable 4×4 multiplier based on full-adder circuits. With shift registers and basic logic functions available along with digital adders (or subtractors) and multipliers, it is obvious that any desired digital function can be synthesized. The question then becomes one of identifying those functions or types of functions which are best suited to CCD digital implementation.

The development of DCCL has been aimed at providing a cost-effective signal processing capability for, primarily, military applications. Signal process-

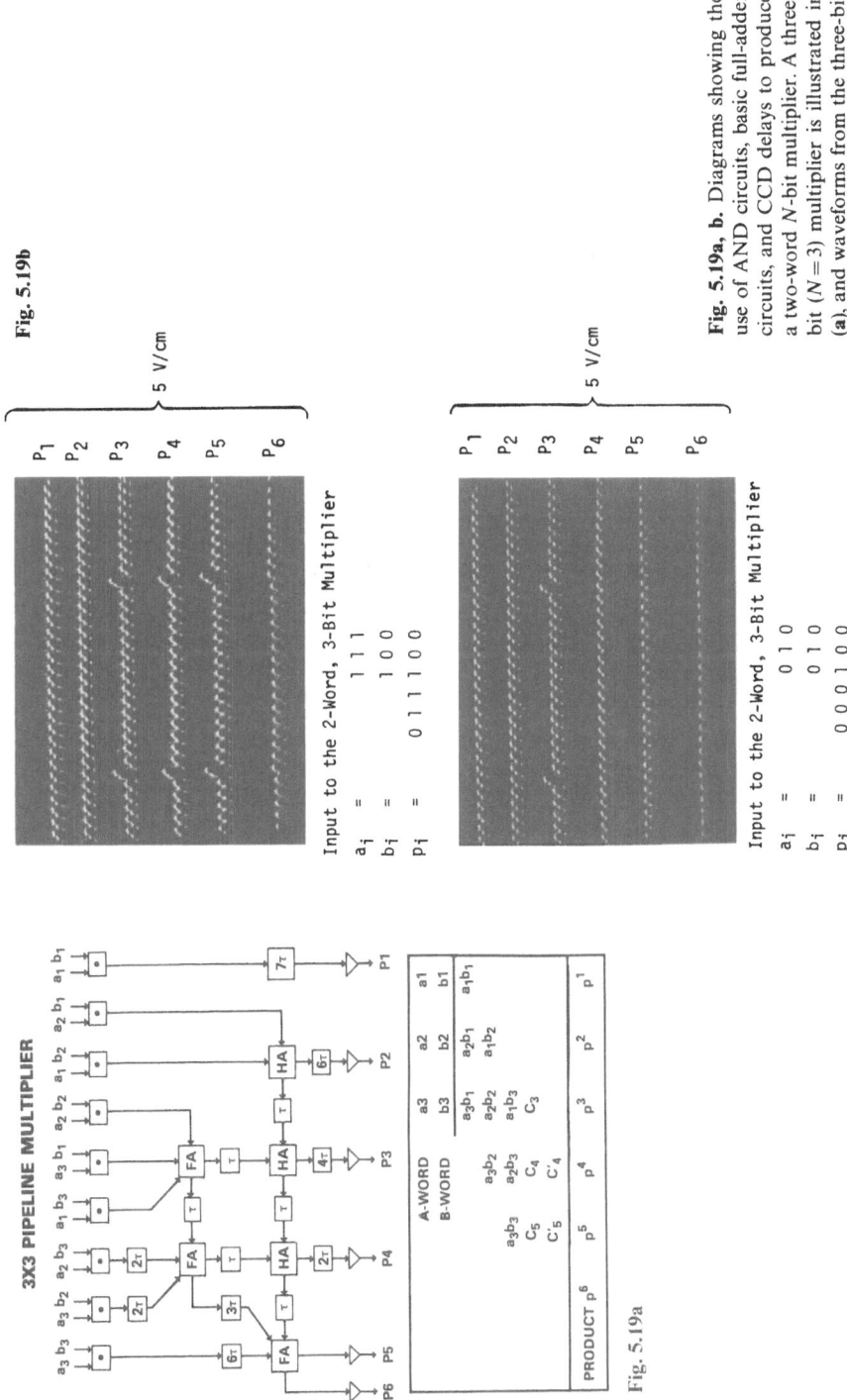

Fig. 5.19b

5 V/cm

Input to the 2-Word, 3-Bit Multiplier

a_i =   1 1 1
b_i =   1 0 0
p_i =   0 1 1 1 0 0

5 V/cm

Input to the 2-Word, 3-Bit Multiplier

a_i =   0 1 0
b_i =   0 1 0
p_i =   0 0 0 1 0 0

Fig. 5.19a, b. Diagrams showing the use of AND circuits, basic full-adder circuits, and CCD delays to produce a two-word N-bit multiplier. A three-bit ($N = 3$) multiplier is illustrated in (**a**), and waveforms from the three-bit multiplier are shown in (**b**) [5.26, 27]

3X3 PIPELINE MULTIPLIER

A-WORD	a1	a2	a3		
B-WORD	b1	b2	b3		
	a_1b_1	a_2b_1	a_3b_1		
		a_1b_2	a_2b_2	a_3b_2	
			a_1b_3	a_2b_3	a_3b_3
			C_3	C_4	C_5
				C_4'	C_5'
PRODUCT p^6	p^1	p^2	p^3	p^4	p^5

Fig. 5.19a

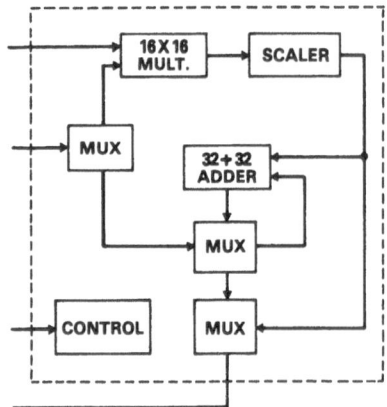

Fig. 5.20. Block diagram of digital CCD arithmetic chip

ing, especially for military applications, often requires a high-performance high-throughput processor which performs highly structured functions. Examples are matched filtering, correlation, convolution, and fast transforms (i.e., Fourier, Hadamard, Hilbert, etc.). The inherent structure to these functions allows them to be cast into a flow form which is ideal for pipeline computations using DCCL. This structure allows timing and data routing to replace much of the program memory and control logic found in general purpose processors [5.28].

A DCCL implementation for realizing flow form signal processing functions can be formed using only two types of DCCL LSI chips. One chip provides the arithmetic functions while the other provides memory and control. Although the accuracy required for different applications can vary widely depending upon the dynamic range, signal-to-noise ratio, expected signal growth due to integration, etc., the chip configurations described below assume 16-bit accuracy with the additional capability of double precision for add/subtract operations. This meets or exceeds the requirements for most applications and provides performance equivalent to most mini-computers.

The basic arithmetic functions to be realized are addition/subtraction, multiplication, and scaling (multiplication by a power of two). The arithmetic accuracy required for the different applications, as we have seen, can vary widely. However, the more stringent applications can be satisfied with 16-bit multiplication accuracy and addition/subtraction accuracy of 16 bits, plus a double precision capability of 32 bits. A DCCL chip configuration with this capability is shown in Fig. 5.20. To allow the sequence of arithmetic operations to be performed in different orders corresponding to different applications, multiplexers are placed at the input to the adder and multiplier. A multiplexer is also provided at the output so that results of the different operations can be selected.

Control inputs are accepted by the arithmetic chip to route the data through the desired elements so that a prescribed sequence of operations is performed. For example, for the FFT "butterfly", the sum and difference of the inputs is formed. The sum is returned to the output for storage while the difference is applied to the multiplier so that the complex multiplication (i.e., polar rotation) can be carried out. To perform a correlation computation, the control inputs to the arithmetic chip cause the output of the adder to be fed back to the input forming an accumulator. The sequences of the two input variables to be correlated are then applied to the multiplier and the product accumulated. Registers for latching the data words are not shown in Fig. 5.20 because storage is implicit in the operation of DCCL.

The fundamental limitation in applying digital charge-coupled logic is the relatively long throughput delay. Best computational efficiency is obtained using pipeline techniques. Although in many of the important applications the computations can be cast into a flow form suitable for pipelining, there is also a need for general-purpose computer operations such as executing branching and jump instructions. Conditional instructions of this type require a comparison to be made before the next step in the program is determined. Due to the long delay through the DCCL arithmetic logic, it is difficult to perform general-purpose computing efficiently. With these characteristics of DCCL arithmetic in mind, it is important to tailor the control chip to match the performance of the arithmetic unit.

The basic timing can be broken up into blocks of N clock intervals where N exactly matches the delay through the arithmetic unit; typically this may be between 16 and 32 clock intervals. Control of the DCCL arithmetic unit can be divided into block and intrablock instructions. The intrablock instruction controls allow pipelined operations to be performed, like the FFT, where data words are added and then subtracted on successive clock pulses. By changing the block instructions, the intrablock instructions can be changed after each block of N clock pulses. For example, the arithmetic unit could perform successive sums and differences on one block of N samples followed by a multiply and accumulate on the next block of N samples. Since the results of an arithmetic operation become available at the arithmetic unit output after N samples, branching, skip, and jump instructions can be performed at the block rate. Thus the arithmetic unit can either perform flow form types of computations at the high rate or general-purpose types of computations at the lower block rate.

The separation of the control of the arithmetic unit into block and intrablock functions suggests a shift register type of architecture for the DCCL control chip as shown in Fig. 5.21. The organization of the chip is not unlike that used for block organized random access memory (BORAM). In this case, however, the blocks of shift register memory may be either conventional data memory of shift register read-only memory (ROM) where a metallization mask determines how many of the blocks are to be ROM as well as the contents of the ROM. Multiplexers at the input and output control the flow of data and program instructions to and from the arithmetic unit(s).

Each memory block in Fig. 5.21 consists of shift registers and recirculation *logic*. Nominally, the block would be 16 or 32 samples long with 16 shift

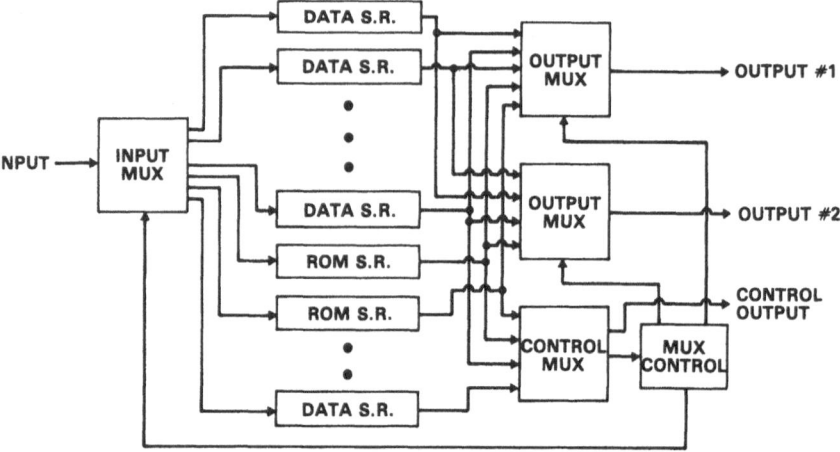

Fig. 5.21. Block diagram of digital CCD controller chip

registers in parallel corresponding to a 16-bit/word parallel data format. As shown in Fig. 5.21, the shift register block has a tap after $N-1$ samples so that the data can be recirculated back to the input after either N or $N-1$ sample delays. This allows the memory blocks to be operated either as recirculating memory or as a delay-line time compressor (DELTIC) so that as the data process through the register, the oldest sample is replaced by the new input sample.

When the DCCL control chip is operated with the DCCL arithmetic unit, the two data outputs are selected by the multiplex gates and applied to the arithmetic unit inputs, while the arithmetic unit output is fed back to the control chip input. The lower multiplex gate on the right in Fig. 5.21 selects the control values from either the ROM shift registers of a data shift register. Thus the controller operation can proceed according to a program stored in the ROM or be changed by input values. The input values can be from either an external source (e.g., an interrupt) or from values computed by the DCCL arithmetic unit.

When operated at the slower block rate, both the data and the next program control values can be time interleaved so that both are computed in a single N clock interval block time. With a 4 MHz clock rate and $N=32$, both operations can be accomplished in 8 μs. This speed, in addition to CCD advantages of high component density and low power, makes the pair of DCCL chips competitive with present microprocessors.

An example of an application of DCCL chips is as follows. In many applications it is necessary to provide a bank of digital comb filters. When the filters are uniformly spaced in frequency, it is convenient to compute the outputs F_k as coefficients of the discrete Fourier transform (DFT)

$$F_k = \sum_{n=0}^{N-1} f_n \exp(-i2\pi nk/N), \tag{5.13}$$

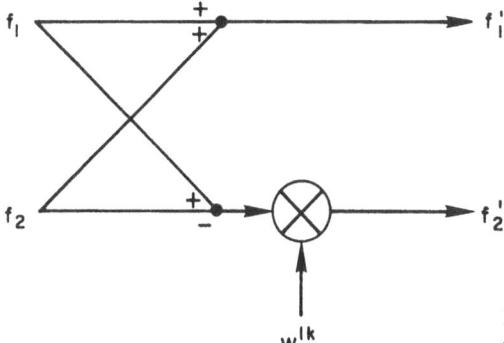

Fig. 5.22. FFT kernel ("butterfly") flow diagram

where the f_n are the set of N input data samples. When the number of filter outputs is less than, say, 16, the F_k are computed directly using the DFT. For larger numbers of comb filters, the well-known fast Fourier transform (FFT) algorithm [5.29] is preferred since the reduction in the number of operations (N^2 for the DFT and $N \log_2 N$ for the FFT) more than offsets the increased programming complexity. Both the DFT and the FFT are easily cast into a flow form which can be computed efficiently with the pair of DCCL chips.

For the DFT it is convenient to compute the inphase (cosine) and quadrature (sine) components of each F_k separately. The N input signal samples are loaded into the data shift register indicated at the top of Fig. 5.21. The shift register ROMs contain the sequence of sine and cosine components corresponding to the DFT reference function, $\exp(-i2\pi nk/N)$. To compute the inphase components of one of the F_k, the data shift register and ROM reference shift register are shifted so that successive signal and reference pairs are applied to the arithmetic chip where they are multiplied and accumulated. The signal samples in the data shift register are recirculated so that at the end of N shifts the computation of the inphase component is complete and the data samples are in the correct position to begin the computation of the quadrature component. During the next N clock pulses the output multiplexer connects the ROM shift register containing the quadrature reference samples to the output. In the arithmetic chip the signal and reference samples are multiplied and accumulated as before. To compute K complex DFT outputs requires $2KN$ shifts. K is usually limited to less than 16, both because the FFT provides better computational efficiency and because the ROM shift register storage of the reference functions becomes a limiting factor.

The FFT computes N complex Fourier coefficients by a sequence of $\log_2 N$ iterations. On each iteration $N/2$ kernel or "butterfly" operations are performed. The flow diagram for the FFT butterfly is shown in Fig. 5.22. Computation of the butterfly consists of six add-subtract operations and four multiples as shown in Fig. 5.23. To perform the FFT requires that the control function provide timing and address information so that the data and the sine and cosine coefficient values can be read out of memory, applied to the butterfly arithmetic, and then returned to memory.

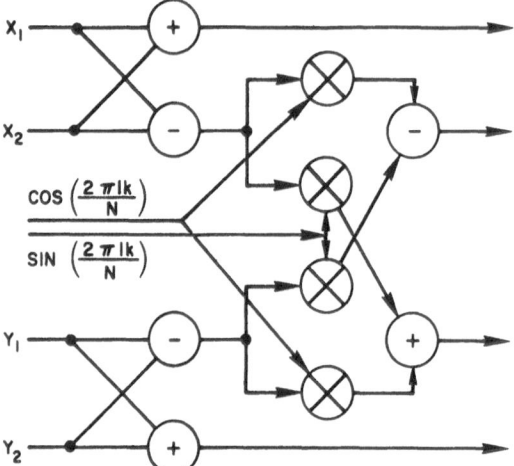

X_1

X_2

$\cos\left(\dfrac{2\pi lk}{N}\right)$

$\sin\left(\dfrac{2\pi lk}{N}\right)$

Y_1

Y_2

Fig. 5.23. FFT kernel operations

The FFT can be computed efficiently with the DCCL arithmetic and controller chip since the FFT can be placed in a flow form suitable for pipeline operation. The technique is based upon a form of the FFT algorithm due to *Singleton* [5.30]. With this technique the data is stored in a pair of serial memories. In our case the serial memories are realized as shift registers in the controller chip. Since the data are complex, four shift registers, each 16 bits wide, are required. Spacing between the data pairs used in the butterfly (i.e., the "wing span" of the butterfly) changes by a factor of 2 on each of the $\log_2 N$ FFT iterations. In order to assure that the correct data pairs appear simultaneously at the shift register output during each FFT iteration, the MUX CONTROL in Fig. 5.20 must cause the data returned to the controller chip input from the arithmetic chip to be leaded into the correct shift register.

The computation of each butterfly can be accomplished either by cyclically passing the data through a single arithmetic unit and performing and add/subtract or multiply on each pass or by employing four arithmetic chips in parallel and performing the complete butterfly computation on one pass. If multiple passes per butterfly are used, then a pair of additional temporary storage shift registers in the control chip is needed.

5.2.3 High-Speed Devices

There is currently growing interest and activity in CCDs which are capable of operating at speeds up to 100 MHz and higher. This large bandwidth has significantly increased the scope of signal processing and systems applications for which CCDs can now be reasonably considered. In this section the most effective configuration for achieving very high speeds, the peristaltic charge-coupled device (PCCD) [5.31, 32], is discussed.

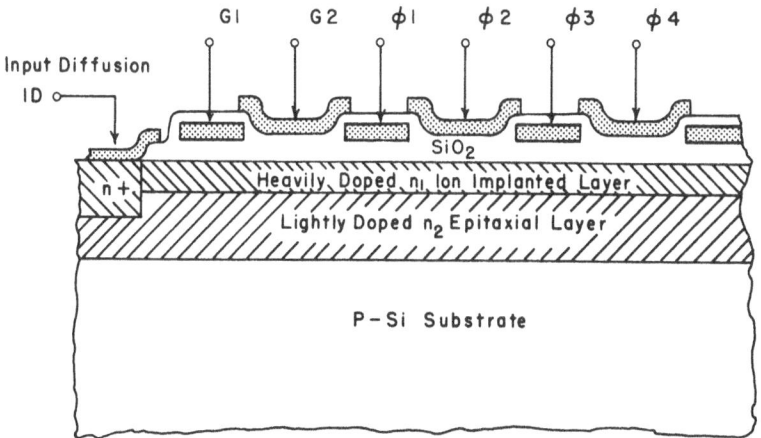

Fig. 5.24. Cross-sectional view of the deep buried-channel (peristaltic) CCD

A four-phase version of this configuration is depicted schematically in Fig. 5.24. The structure consists of a p substrate with a thick ($\sim 5\,\mu$m) high-resistivity (doping level $\sim 5 \times 10^{14}$/cm^3) epitaxial layer on which a thin ($\sim 0.3\,\mu$m) n^+ layer is implanted (dosage $\sim 10^{12}$/cm^2). The gates (either aluminum or polysilicon) are isolated from the n layer by a thin (1000–2000 Å) oxide layer. There are efforts underway to replace the epitaxial layer with an ion-implanted layer in which the impurities are diffused deeply into the substrate. This double-implanted structure is expected to have several advantages over the epitaxial structure in Fig. 5.24, associated with a lower defect density in the CCD channel, including: lower leakage current density and leakage nonuniformities, lower bulk charge trap densities, better control of the doping, and higher yield. There are two important features of the PCCD structure which must be noted. First, the n layer is very thick. This gives rise to efficient charge transfer at high speeds due to the large transverse fringing fields. Second, the n^+ layer permits a large charge handling capability.

The primary reason for the high speed of the PCCD is the thick bulk n layer. Charge is carried in the bulk, away from the surface traps as in conventional buried channel structures. In addition, the very thick n layer gives rise to an electric field profile which is favorable for high-speed operation. At the beginning of a charge transfer in all CCD structures the high field at the edge of the gates forces carriers in that region quickly into the next cell. This causes a charge gradient to form under the gate of the full cell. The resulting self-induced transverse field is the dominant mechanism for the initial stage of charge transfer, but since it is due to the mutual repulsion of charge carriers it is important only when the charge density is high. In surface-channel and shallow buried-channel CCDs diffusion is the primary mechanism for transfer of the electrons remaining after the self-induced transfer. Thus, it is the last relatively few electrons which limit the transfer efficiency at high speeds. In the PCCD these last few carriers are driven deep into the bulk by the external field from the gate. Since the epitaxial layer thickness is comparable to the gate length, transverse electric fields deep in the layer are very large. This high field pushes the electrons rapidly into the next cell. Thus charge transfer in the PCCD structure is strongly field enhanced.

Theoretical calculations [5.33] and experiments using a uniphase clock [5.3] have projected operating data rates of the order of 1 GHz for the PCCD with charge transfer inefficiencies of around 10^{-4}. These operating speeds have not, however, been achieved in practice because presently available input, output, and driver circuits do not operate fast enough. Therefore the major part of high-speed CCD research involves configuring clocks, inputs, and outputs that are capable of high-speed operation.

The most common clocking technique is the four-phase arrangement shown in Fig. 5.24, which has been used to clock deep buried-channel (peristaltic) CCDs faster than 100 MHz [5.2, 33]. The same gate configuration can also be used with a two-phase clock by offsetting the potential on the ϕ_2 and ϕ_4 gates with either a dc bias or an additional ion implant. Adjacent gates are then connected and driven together (i.e., the ϕ_1 and ϕ_2 gates are driven with one phase; the ϕ_3 and ϕ_4 gates are driven with the other phase). In comparison with the four-phase driver, this offers simpler logic circuitry plus twice the data rate for the same driver frequency at the expense of some linearity and charge handling. Another consideration is clock power dissipation, which is particularly important at high frequencies. If the device is intended for continuous operation at a single frequency, a great deal of power can be saved by designing the driver circuit with inductance that resonates the capacitive reactance of the CCD gates. However, changing frequencies with resonant drivers is a problem.

A signal voltage can be converted to a charge packet at the CCD input by a number of methods. The most linear lowest noise input technique is known as fill and spill, which can be explained using the configuration shown in Fig. 5.24. The input signal voltage is applied to the storage gate G2 and the input diffusion ID is pulsed to a potential which causes charge to flow through G1 into the G2 potential well. The ϕ_1 potential blocks further charge flow during this time. When the G2 well is filled ID is returned to an attractive potential, so that excess charge spills back through G1 into the diffused region until there is no charge stored under G1. After the charge flow is completed, a precisely metered charge packet remains under the gate G2. The quantity of charge in the packet is directly proportional to the voltage difference between G1 and G2. The charge packet is transferred out of G2 when ϕ_1 is clocked to form a potential well. The noise associated with the fill and spill input technique is due to thermal fluctuations and is thus very low. Shot noise is not present since the charge packet is allowed to equilibrate. Unfortunately, this equilibration requires times of the order of a few ns since its speed is limited by thermal diffusion. This rules out the use of fill and spill for frequencies above about 100 MHz, and the linearity may not be adequate for some applications at frequencies as low as 10 MHz.

Another input for CCDs, which is a charge partitioning method, can also be explained using the configuration shown in Fig. 5.24, although a third gate is typically included between the input diffusion and G1 to act as a buffer. In this input technique the signal is applied to either one of the terminals ID or G2, while the other is held at a fixed potential. G1 is pulsed to allow charge to flow freely from ID to the G2 potential well. Further charge flow is blocked by ϕ_1 during this time. When the well is filled the quantity of charge under the G2 gate is proportional to the voltage difference between ID and G2. The voltage on G1 is then abruptly changed to form a potential barrier which isolates the charge packet under G2. Applying the input signal to ID rather than a high-impedance gate has the advantage of making the technique more compatible with low-impedance sources. On the other hand, the sampling linearity is better when G2 is used for the input terminal.

Several characteristics of the charge partitioning input improve its high-speed performance. As the input signal is sampled, the area under G1 is in full conduction, allowing the charge to reach equilibrium more quickly. When the gate G1 is closed to partition the charge, the high transverse fields associated with the deep buried-channel configuration strongly enhance charge motion so that the charge packet is isolated very rapidly. The charge partitioning input is satisfactory for frequencies as high as a few hundred MHz.

An even faster input circuit has been designed using bipolar transistors [3.34, 35]. The circuit consists of a common base double-diffused bipolar mirror structure. In this structure the collector of the bipolar transistor is part of the CCD channel, i.e., the CCD holding well. Current injected into the emitter will enter the collector through the narrow base region. Since the base can be made less than $1 \mu m$, the charge enters the holding well very quickly. Special clocking or gating pulses are not necessary in this input technique; however the amount of charge injected into the CCD holding well depends not only on the magnitude of input current but on the clock dwell time. Thus this bipolar transistor input method is most effective for applications where the high-speed sampling and clocking takes place at a fixed rate. An additional speed advantage results from configuring the transistors in a mirror structure because the effective parasitic capacitance is reduced by the ratio of emitter-base junction areas of the bipolar input transistor and its mirror. Calculations have been made which predict linear operation of this input circuit at frequencies approaching 1 GHz.

The very high-speed input techniques, both charge partitioning and bipolar transistor mirror, cause a charge packet to be created very rapidly without thermal equilibrium. Since thermal equilibrium is not reached there will be shot noise added to the input signal. The shot noise charge is equal to the square root of the injected signal charge, so that for a given dynamic range this places a limit on the minimum allowable signal level.

There are several techniques for taking serial output signals from CCDs, none of which have been very effective at very high speeds. At present, the output circuitry represents the most severe limitation on CCD operating speed.

Future improvements will probably include: using very high-frequency FETs to perform the on-chip amplification, using bipolar amplifiers because of their high speed and low impedance output, or, perhaps for some applications it may be possible to remove the signal directly from the output diffusion and amplify off-chip.

It can be concluded that CCDs capable of operating at hundreds of MHz are currently achievable but are presently limited by peripheral circuitry.

A new configuration which has a great deal of potential for very high-speed operation is the gallium arsenide CCD [5.36]. Charge transfer efficiency in excess of 0.999 has been measured on a device with Schottky barrier gates. This gate structure circumvents the problems of high surface-state density and instabilities which have plagued efforts to form effective gate dielectrics on GaAs. The substrate configuration consists of an *n*-type active layer approximately 2 μm thick on a semi-insulating substrate. The potential profile can be designed for deep buried-channel operation, with high transverse fields to enhance high-speed performance. There are several reasons why GaAs is attractive for high-speed CCDs. The low field electron mobility is 4 to 5 times higher than in silicon, so that efficient charge transfer at higher speeds should be possible. Schottky gate GaAs FETs capable of operating at microwave frequencies [5.37] are process compatible with the CCDs and could be used for output amplifiers and on-chip clock drivers. High-speed GaAs logic technology [5.38] could be used for input sampling and clock circuitry. Thus GaAs-based CCDs offer an attractive approach to very high-speed analog signal processing, although much work remains to be done in all aspects of this relatively new technology.

5.2.4 CCD-SAW Devices

This section contains a description of surface acoustic wave (SAW) [5.38a] linear FM (chirp) filters, their applications for real-time analog Fourier transformation with the chirp-*z* algorithm, and the importance of combining these devices with CCDs. A few years ago it was thought that CCDs and SAWs would compete strongly for many analog signal processing applications. More recently a number of authors comparing CCDs and SAWs have emphasized the difference between the technologies in order to show they are complementary rather than competitive. Even more recently, since each technology has established a rather solid niche in signal processing, people have begun to rethink the properties of CCDs and SAWs with a eye toward combining the two in order to perform sophisticated signal processing functions. Although various interesting ideas have been proposed, the combination of a CCD delay-line time compressor/expander with a SAW chirp-*z* transformer has the most obvious importance for signal processing. Discussion of this combination follows a description of the SAW device.

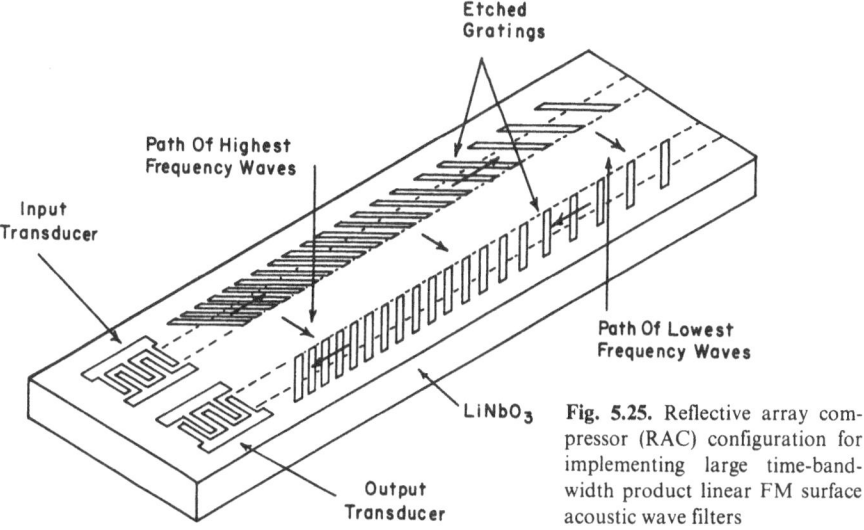

Fig. 5.25. Reflective array compressor (RAC) configuration for implementing large time-bandwidth product linear FM surface acoustic wave filters

Large time-bandwidth product linear FM SAW filters are usually made using the reflective array compressor (RAC) configuration [5.39], shown schematically in Fig. 5.25. A surface wave is excited by applying an rf signal to the broadband input interdigital transducer. The wave propagates into the array of shallow grooves where it is reflected into the second row of grooves and reflected again toward the output transducer. There are typically thousands of grooves formed by ion milling or sputter etching on a single device. One groove by itself reflects very little acoustic power but the cumulative effect of a large array becomes a very efficient surface wave reflector when the periodic spacing of the grooves is equal to one acoustic wavelength. For a linear FM filter the groove periodicity varies linearly with position along the array so that the highest frequency waves are reflected at the portion of the array nearest the transducers, while the lowest frequency waves propagate to the far end of the array before being reflected.

Some state-of-the-art performance achievements for SAW dispersive filters are listed here. The highest time-bandwidth product obtained is 16,200 using a time delay of 90 μs and bandwidth of 180 MHz [5.40]. Time-bandwidth products as large as 10^4 are very difficult to achieve, however, so perhaps a more practical upper limit is a few thousand with present state of the art. Dispersive bandwidths of 500 MHz have been demonstrated by a number of companies and this number will be surpassed, at least by a factor of 2, as the inevitable demand for higher bandwidths arises. However, bandwidths below 200–300 MHz are much more routine and below 100 MHz are generally easy.

The maximum dispersive delay time is of course the length of time required for a wave to propagate from one end of the groove array to the other end and back. The surface wave velocity of the most commonly used substrate, lithium

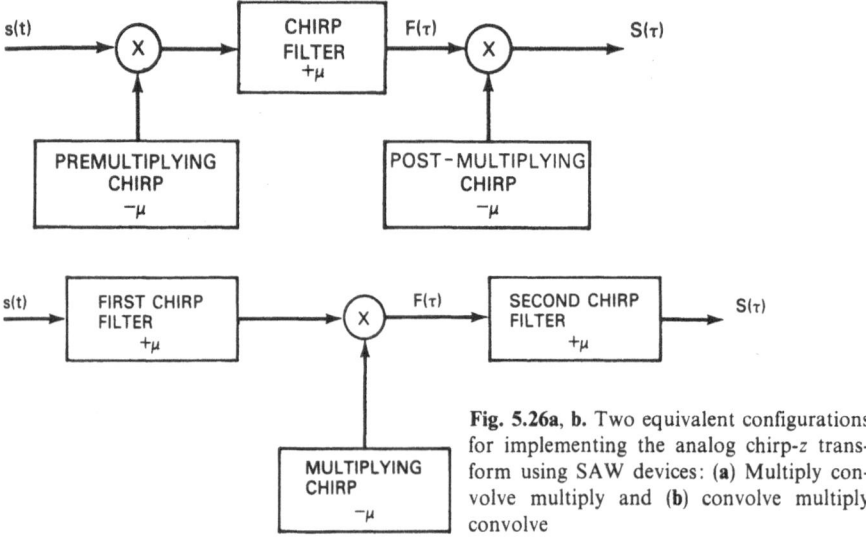

Fig. 5.26a, b. Two equivalent configurations for implementing the analog chirp-z transform using SAW devices: (a) Multiply convolve multiply and (b) convolve multiply convolve

niobate, is 3.49×10^5 cm/s so that $100\,\mu s$ dispersive delay requires parallel groove arrays at least 17.5 cm long. In spite of the obvious difficulties in making such a device, lithium niobate crystals of sufficient size are available and special fabrication techniques have been developed so that $100\,\mu s$ is generally considered to be a practical upper limit. Longer times could be achieved using substrates with slower surface wave velocity, such as bismuth germanium oxide [5.41], but such crystals are not available with size and quality comparable to lithium niobate.

One important application for the SAW chirp filter is the chirp-z transform configurarion [5.42–44] shown in Fig. 5.26. The two algorithms shown are duals of one another and perform the same function, although the Fig. 5.26a algorithm has been more popular because the signal is required to pass through only one dispersive filter. In this configuration the signal is premultiplied, using a mixer with high linearity and dynamic range, by a chirp with negative slope μ, then passed through a chirp filter with positive slope μ, and finally post-multiplied by a chirp with negative slope μ. If only the power spectrum is desired the post-multiplication can be omitted. The multiplying chirps could either be generated actively using a swept oscillator, or by impulsing SAW chirp filters.

It is important to remember that all the chirps in the processor have finite length and bandwidth. This, of course, limits the time duration and the bandwidth of input signals which can be transformed. It can be shown [5.42] that if the dispersive filter bandwidth is B and its length is T, then the maximum input signal time-bandwidth product for the true Fourier transform is $TB/4$, corresponding to a signal bandwidth and length of $B/2$ and $T/2$, respectively. Thus for a 1000-point Fourier transform, the time-bandwidth product of the dispersive filter must be at least 4000.

Clearly, Fourier transforms with 10^3 points can be achieved in a few tens of microseconds using the SAW CZT algorithm. This real-time performance is far beyond the speed of the current digital FFT. The SAW implementation is also more attractive than the CCD CZT for many applications, particularly when high speed and high time bandwidth are required. Even if CCDs eventually match the high-speed and time-bandwidth performance obtained with surface waves, the SAW still offers a hardware simplification since it is an if rather than a base-band processor. The single chirp filter shown in Fig. 5.26a would have to be replaced by four filters in a base-band system [5.45] (a sine and a cosine filter for the I and the Q channels, as shown in Fig. 5.4).

A significant drawback of the SAW CZT processor is that once the device has been fabricated its bandwidth and time delay are fixed. It is capable of very high processing speeds but its input and output must also be very fast. This presents an interfacing problem in many potential systems applications. CCDs, with their variable clock speed, will probably have an important impact in this area. A CCD time-base compressor could be used to receive signals with relatively low bandwidths, then speed up the clock to convert the signals to a higher bandwidth compatible with a SAW device. Similarly, a CCD time-base expander could be used at the output of a SAW device to slow down the SAW output so that it is more easily compatible with subsequent processors. A CCD analog storage array at the SAW input can also be used to multiplex many low-frequency signals so they can be processed sequentially by a single fast SAW device. One example of this is the CCD-SAW sonar beam forming technique proposed in [5.46].

It is evident that for the CCD-SAW combination to be most effective, development of high-speed CCDs must be continued. For example, a 500-point SAW chirp-z transformer requires a dispersive filter time bandwidth of at least 2000, as discussed earlier. Since 100 μs is the maximum reasonable dispersive time delay, the minimum filter bandwidth is 20 MHz, which corresponds to a minimum signal bandwidth of 10 MHz. Since the CCD is a base-band sampler it must load samples into the SAW at a 20 MHz rate. This is already fast for currently available CCDs but in order to take maximum advantage of the high speed of SAW devices even faster CCDs should be developed.

By using the variable clock feature of CCDs to time expand or compress signals to fit the requirements of SAW devices it is possible to take advantage of both the flexibility of CCDs and the signal processing capabilities of SAWs. Thus the CCD-SAW combination could be useful for signal-processing applications not previously considered for either technology alone.

5.3 Applications

The charge-coupled device structures discussed in this chapter have a wide variety of applications. Among the signal-processing application areas where

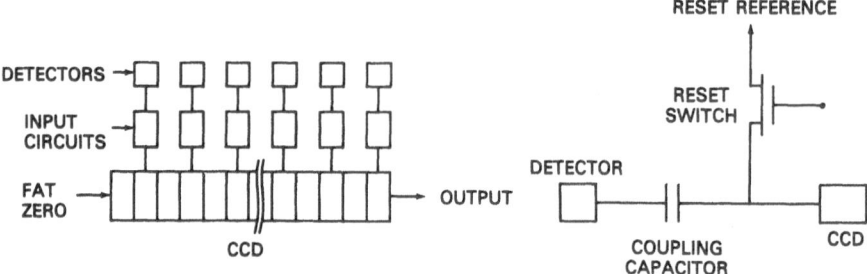

Fig. 5.27. CCD multiplexer organization for detectors   **Fig. 5.28.** Simplified ac coupling technique for CCD multiplexer input

CCDs are having substantial impact are electrooptical systems, sonar, voice communications, and radar. In this section, these applications and the ways in which CCDs can be used are discussed.

5.3.1 Electrooptical Systems

Charge-coupled devices have important applications in signal processing for electrooptical systems, including visible and infrared. They are especially significant in efforts to develop very high-density, high-performance focal plane arrays.

Detector Readout
The CCD is useful as a multiplexer for detectors in several configurations. Figure 5.27 illustrates the concept. The detectors may be discrete detectors made on a separate substrate from the CCD, resulting in a hybrid array. Alternatively, the detectors may be made on the same substrate as the CCD, resulting in a monolithic array. Present technology tends to limit high-performance CCDs to silicon, with the result that for infrared systems the hybrid arrays are somewhat more advanced than the monolithic arrays. The hybrid approach using the CCD as a multiplexer has been applied to mosaic (staring) [5.47–49] and scanned infrared arrays [5.49]. Monolithic CCD multiplexers are used in silicon visible imagers [5.50] and research is underway to develop CCD technology in HgCdTe and InSb (see, for example, [5.51]) so that monolithic infrared focal planes can be fabricated. In all multiplexer applications, the interface circuit between the detector and the CCD is of primary importance. The nature of the interface circuit depends on the type of detector (PC, PV, MIS, pyroelectric, etc.), on the detector bandwidth required, and on such other system requirements as power dissipation and detector pitch. In addition, the CCD questions of different offsets from input to input and of cross talk due to transfer inefficiency in the multiplexer may have to be considered. Estimates have been made that the threshold voltage of devices a few mils apart on a silicon substrate may vary as much as 100 mV due to

substrate resistivity nonuniformities [5.52]. This problem may be more severe in other substrate materials. Thresholds variations combined with variations in detector offsets and responsivity may require ac coupling between the detectors and the CCD multiplexer, especially for infrared applications. A simple ac coupling scheme of a type easily incorporated on the CCD is shown in Fig. 5.28. A novel alternative input approach for dealing with spatial variations in threshold voltages has been devised which to first order is insensitive to threshold voltage [5.52]. Cross talk can arise in a multiplexer as a result of transfer inefficiency with charge from one channel being trapped and then reemitted into the charge packet of another channel. If a charge packet Q is introduced into a CCD well at position $i=0$, then after N stages of transfer the distribution of charge in the cells will be given by

$$Q_i = Q(1-\alpha)^i \alpha^{N-i} N!/(N-i)!i!, \tag{5.14}$$

where α is the fractional charge loss per stage. The cross talk between adjacent channels ($i=N$ and $i=N-1$ for the worst case) after N stages of transfer can be written as

$$Q_{N-1}/Q_N = N\alpha/(1-\alpha). \tag{5.15}$$

For a 50-stage multiplexer with $\alpha = 10^{-4}$, the cross talk is about $-46\,\text{dB}$ from (5.15). A 40-stage multiplexer with ac coupled inputs has been reported with an ac dynamic range of 50 dB, a worst-case cross talk of $-44\,\text{dB}$, and an operating frequency of 4 MHz [5.53].

For some infrared applications, the CCD would have to operate at low temperatures. Devices which perform satisfactorily at room temperature generally operate equally well (or perhaps a little better) down to 77 K. In addition, operation of a deep buried-channel CCD at 20 K with a clock frequency of 15 MHz has been demonstrated [5.45].

An extension of the CCD-multiplexer concept is the use of the CCD to perform time-delay and-integration (TDI) processing for mechanically scanned detector arrays. Figure 5.29 illustrates this approach in which the CCD is clocked at a rate which corresponds to the motion of the scene from detector to detector. As a result, the signal from the scene is added coherently at each of TDI. If the noise sources are random, the noise will add incoherently at each of TDI and a signal-to-noise improvement equal to the square root of the number of TDI stages can be achieved. This technique has been applied to a visible imager of 128 lines with 128 TDI stages per line and an output multiplexer for a study of day/night periscope feasibility [5.55]. High-density infrared arrays on extrinsic silicon have been fabricated which incorporate CCD TDI registers and multiplexing (see, for example, [5.56]), as well as hybrid arrays in which discrete ir detectors are connected to the silicon CCD TDI [5.57]. An infrared focal plane concept using an InSb charge-injection device (CID) array as a *detector array with silicon CCD TDI processing has been described* [5.58, 58a]

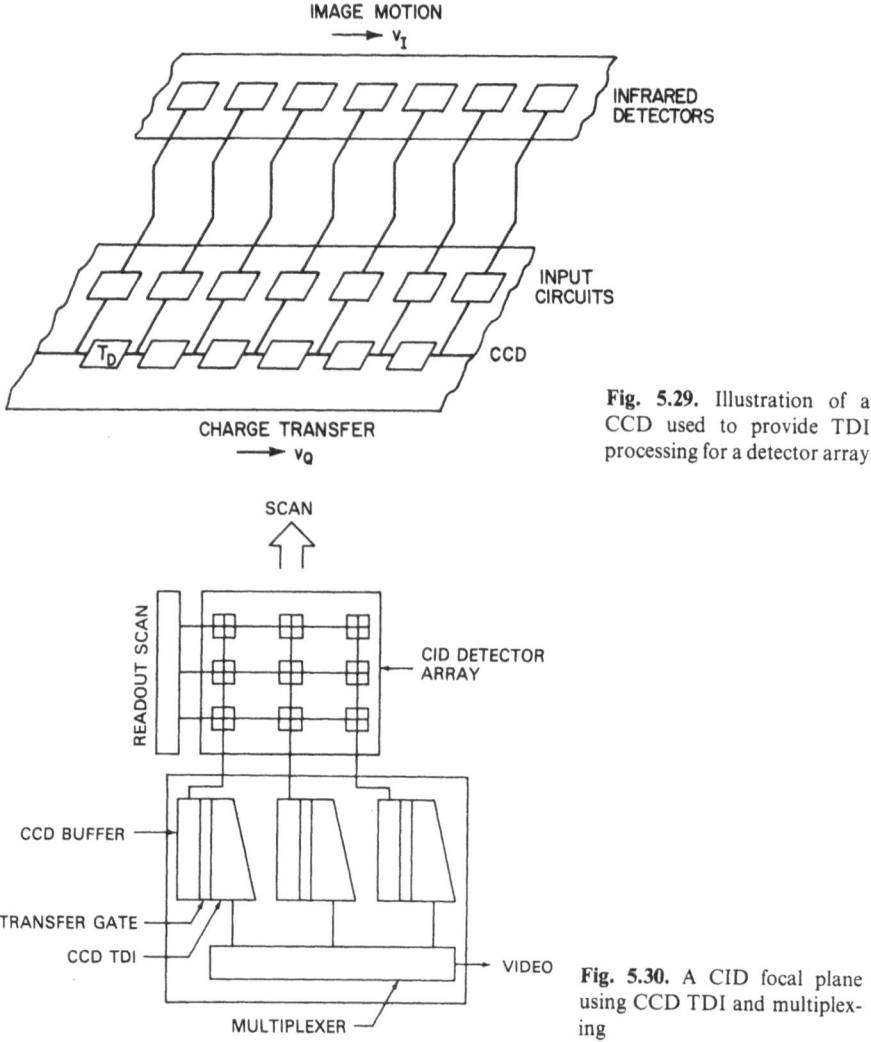

Fig. 5.29. Illustration of a CCD used to provide TDI processing for a detector array

Fig. 5.30. A CID focal plane using CCD TDI and multiplexing

and is under development. Figure 5.30 illustrates this concept. The detector columns are read out into CCD buffer registers from which the data are transferred in parallel into CCD TDI registers. The data in the TDI register are stepped one stage and the readout sequence is repeated. A CCD multiplexes the outputs of the TDI registers into a single video channel.

The problems associated with input and output circuits for TDI are the same as for multiplexers. An additional problem for TDI is signal handling capacity since signals from several detectors are being summed in the CCD. This problem can be especially troublesome in infrared arrays due to the large background signals. One solution is to taper the CCD so that a larger geometry channel is available at the output end (note Fig. 5.30). In low-contrast visible or

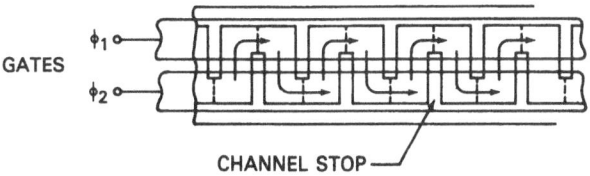

Fig. 5.31. Illustration of a meander channel CCD [5.59] for high-density applications

in infrared imaging, it may be possible to avoid tapering the registers by using ac coupling to the inputs. For high-density focal planes, an important consideration is the physical width of the TDI CCD since that can limit the minimum detector spacing on the array. A device structure which achieves improved packing density has been described [5.59]. This so-called meander channel CCD, shown in Fig. 5.31, reduces the amount of space required for clock lines and their connections.

Signal Filtering

The CCD can be used to perform a number of filtering functions which have potential application to electrooptical systems. As mentioned previously, CCD delay lines can be used to realize recursive filters and filter banks [5.60]. Most applications of CCDs to filtering have concentrated on their use as transversal filters. Figure 5.32 illustrates how PI/SO and SI/PO devices can be used to realize canonical transversal filters. The PI/SO device has the advantage that summation is performed in the CCD itself and input connections to the outside world probably would not require much buffering. The SI/PO device allows for somewhat more straightforward tap weighting. A particularly attractive form of the SI/PO transversal filter is the so-called split-electrode filter described previously (see Fig. 5.3 for an illustration of the split-electrode filter). This approach is well suited for spectral filtering (e.g., low pass filters) as well as for performing transforms. Linear phase filters can be readily constructed, and filter response can be adjusted by changing clock rates.

Scan Conversion

The ability of a CCD to serve as an analog memory suggests applications as scan converters. For visible imaging at very low frame rates, either for very long integration times or for reduced readout bandwidth, CCD analog frame storage can be used [5.61]. In infrared systems, conversion from horizontally scanned detector data to conventional TV presentation can be accomplished. A simplified version of such a converter is shown in Fig. 5.33. In this system, the detector signals are clocked into the CCD and stored during the scan time. The signals are then read out for display one line at a time during the mechanical scan retrace.

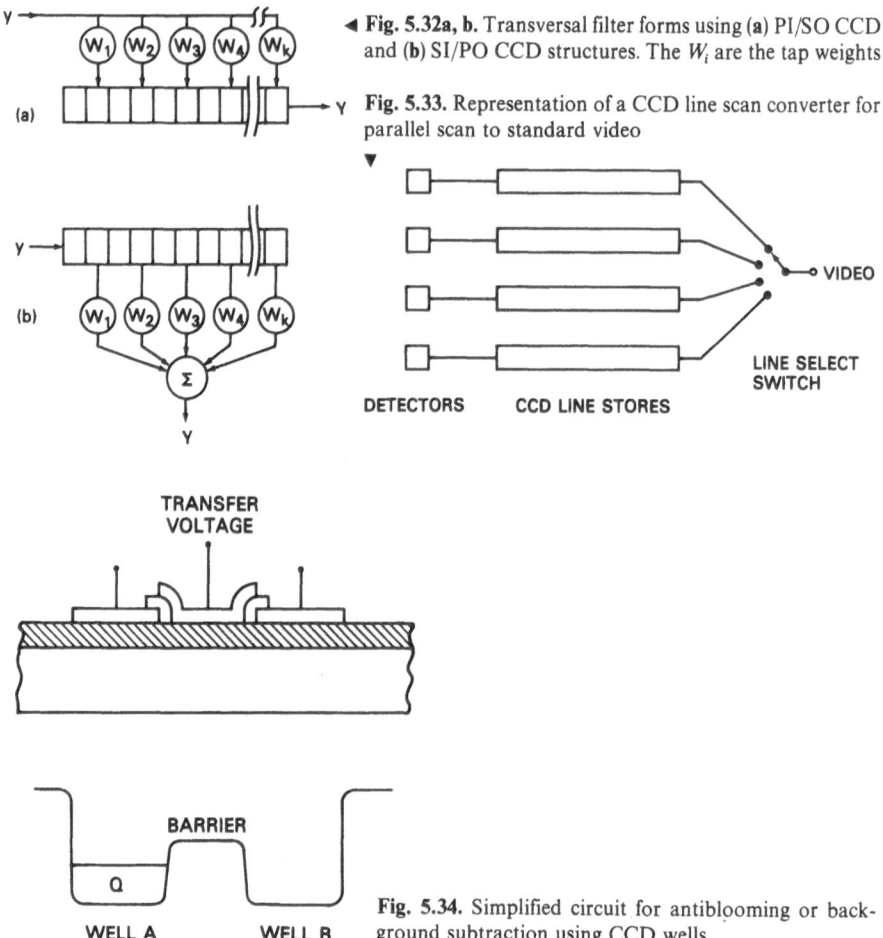

◄ **Fig. 5.32a, b.** Transversal filter forms using **(a)** PI/SO CCD and **(b)** SI/PO CCD structures. The W_i are the tap weights

Fig. 5.33. Representation of a CCD line scan converter for parallel scan to standard video

Fig. 5.34. Simplified circuit for antiblooming or background subtraction using CCD wells

Background Control

A CCD structure can be used to provide antiblooming protection or background skimming in an array. A simple three-gate scheme to illustrate the concept is shown in Fig. 5.34. For antiblooming, the transfer electrode is used to set the potential barrier so that any charge accumulating in well A in excess of a saturation signal will flow over the barrier into B which is periodically emptied into a charge drain. For background subtraction, the charge is accumulated in well A with the transfer gate set so that the top of the charge packet is "skimmed" into well B. The charge in well B represents the signal, while the background charge in well A is periodically dumped to a drain. In the case of antiblooming, the principal problem is the extra device chip area required for implementation. In background subtraction, a major problem is that threshold variations from site to site tend to introduce large pattern noises.

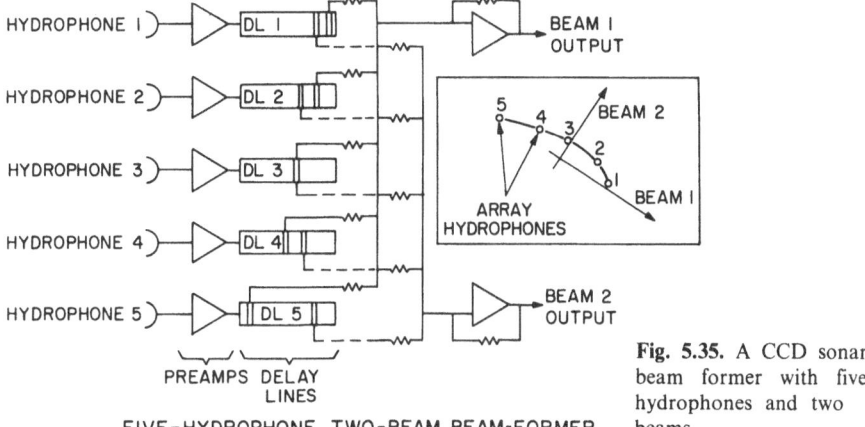

HYDROPHONE 1

HYDROPHONE 2

HYDROPHONE 3

HYDROPHONE 4

HYDROPHONE 5

PREAMPS   DELAY
LINES

DL 1

DL 2

DL 3

DL 4

DL 5

BEAM 1
OUTPUT

BEAM 2
OUTPUT

ARRAY
HYDROPHONES

BEAM 2

BEAM 1

FIVE-HYDROPHONE, TWO-BEAM BEAM-FORMER

Fig. 5.35. A CCD sonar beam former with five hydrophones and two beams.

5.3.2 Sonar

Charge-coupled devices also have potential applications in sonar systems. The use of recursive and/or transversal CCD filters for spectral analysis is directly applicable to sonar, especially for systems which have constraints on size, weight, and power. Another interesting sonar application involves beam forming and steering. In this case the CCD provides the appropriate delay in each hydrophone channel to form the beam or beams from the hydrophone array. A pictorial illustration of this is shown in Fig. 5.35 using SI/PO devices. Beam formers can also be built using PI/SO structures [5.62]. Adaptive signal processing elements can also be realized using CCDs. One approach is to use a SI/PO device as a transversal filter with electrically alterable analog tap weights described in Sect. 5.2.1. These tap weights can then be adjusted under the control of an optimization algorithm. One possibility for the alterable tap weights in MNOS memory transistors used as variable conductances [5.63]. If sufficient uniformity in MNOS transistor characteristics can be obtained, these adaptive filters could fill numerous systems needs [5.63a].

5.3.3 Voice Communications

In government and industrial voice communications there is often a need for privacy. Similarly in military voice communications there is often a require-ment for security. Presently there is considerable interest in the development of secure voice communications systems which can be used with standard telephone and if links. This requires the extraction and encryption of the key parameters which represent the speech samples. These data are usually transmitted over communications links having barely enough bandwidth to pass the original speech; therefore, considerable data compression (perhaps 15:1) is required. The data are then transmitted and reconstructed into speech

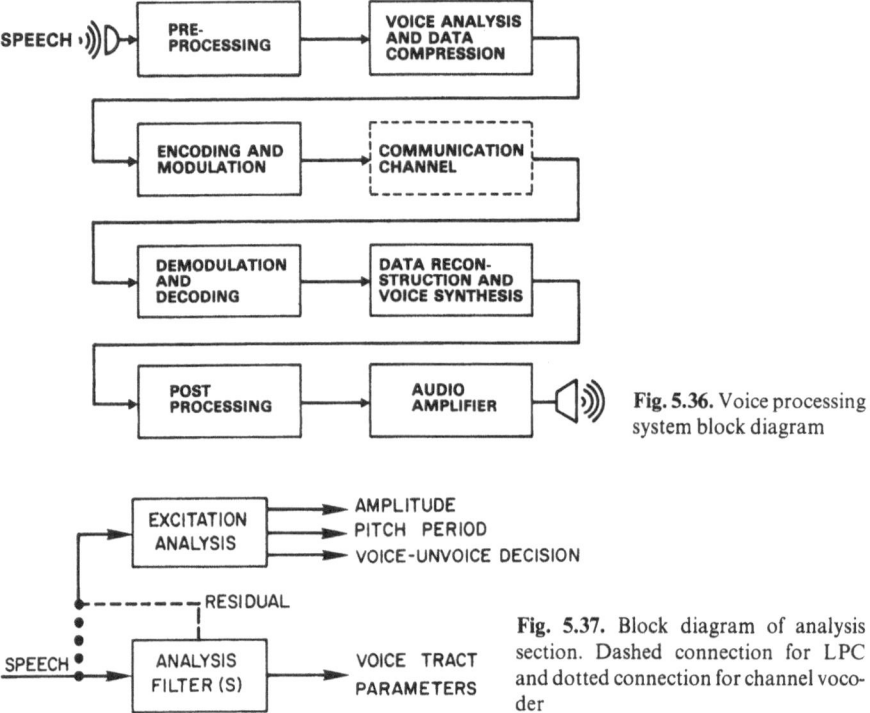

Fig. 5.36. Voice processing system block diagram

Fig. 5.37. Block diagram of analysis section. Dashed connection for LPC and dotted connection for channel vocoder

at the receiving terminal. The transmitted "speech" parameters control the characteristics of the speech synthesizer at the receiver. The excitation is a pulse train for voiced sounds (i.e., vowels) and random noise for unvoiced sounds (i.e., consonants). The two main algorithms which are used in these systems are the linear predictive coder (LPC) and the channel vocoder. The LPC attempts to preserve the time waveform of the speech [5.64].

The basic block diagram of a voice processing system is shown in Fig. 5.36. The transmitter consists of preprocessing, voice analysis and signal compression, and encoding and modulation. The receiver section consists of demodulation and decoding, signal reconstruction and voice synthesis, and postprocessing. The communications channel between the transmitter and receiver can be a telephone link or high-frequency radio. The preprocessor consists of an audio filter (and an A/D converter for digital systems). The post-processor consists of an audio filter (and a D/A converter for digital systems). Figure 5.37 shows the block diagram of the voice analysis and signal compression functions.

For some LPC systems, this consists of the serial combination (dashed line in Fig. 5.37) of an analysis filter and an excitation analysis circuit. The analysis filter provides the vocal tract parameters, i.e., a set of slowly varying predictive coefficients which represent the resonance characteristics of the vocal tract. The wide-band output of the analysis filter is called the prediction residual which is the difference between the actual and the predicted voice samples. The residual is then further analyzed into three parameters: 1) the voice/unvoice decision, 2) the excitation power level (amplitude), and 3) the pitch period (if voiced). A linear analysis based on Gauss' least squares method leads to an adaptive transversal filter as the analysis filter hardware component. Iterative methods can also be used in the analysis. An iterative analysis originated by *Itakura* and *Saito* [5.65] leads to the so-called Ikatura filter as the analysis filter hardware component.

For a channel vocoder system, the analysis filters and the excitation analysis circuit are connected in parallel (dotted line in Fig. 5.37). The excitation analysis is achieved by a bank (typically 19) of narrow-band analog filters. An excitation analysis method presently receiving attention which is useful with the channel vocoder is the homomorphic cepstral pitch detector [5.66]. Figure 5.38 shows the homomorphic cepstral unit in more detail. In the cepstral unit the speech undergoes a Fourier transform, a log-magnitude operation and an inverse Fourier transform. The output contains both vocal tract resonance information and excitation parameters. However, in this chapter, the cepstral unit will only be considered for obtaining the excitation parameters. An impulse appears at the output of the cepstral unit if the speech is voiced. The voice/unvoice decision and pitch period can be obtained from the presence/absence and position of this impulse.

The synthesizer section, shown in Fig. 5.39, contains a variable period pulse-train generator, a noise source, a switch, a variable gain amplifier, and a synthesis filter. The vocal tract parameters control the characteristics of the filter(s). The amplitude or power parameter controls the gain of the amplifier, the pitch-period parameter controls the period of the pulse train generator, and the voice/unvoice parameter controls the position of the switch to connect either the pulse generator or the noise source to the synthesis filter. The output of the synthesizer should be an intelligible reproduction of the original speech. High-performance voice processing systems have been designed and built. However, it is desirable to reduce the size, weight, cost, and power dissipation.

The use of split-electrode transversal filters for the narrow-band analysis and synthesis filter banks in the channel vocoder is obvious. It is estimated that all 19 filter channels could be fabricated on a single CCD integrated circuit chip. Thus only one chip would be required for each of the analyzer and synthesizer filter banks. The primary emphasis in this approach to implementation of voice processing systems is low cost. The pitch tracker would be fabricated on a separate chip. Another use of the split-electrode transversal filter is to perform the chirp-z transform and inverse transform required by the homomorphic deconvolution voice processing algorithm.

The principal use of DCCL in secure voice processing is in the implementation of the Itakura algorithm for analysis and synthesis. The procedure for the analyzer part of the voice processor consists of passing the signal through *ten identical stages*. At each stage a partial correlation coefficient is determined.

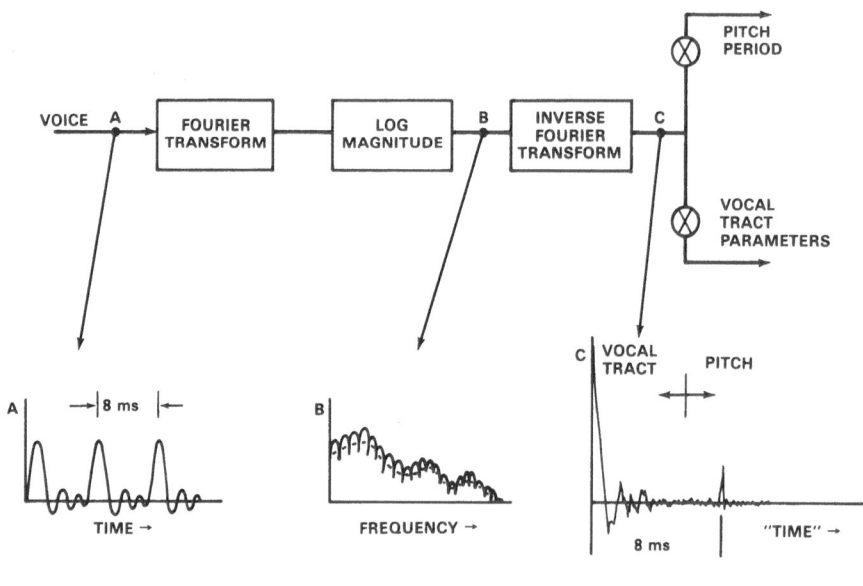

Fig. 5.38. Block diagram of the homomorphic cepstral unit

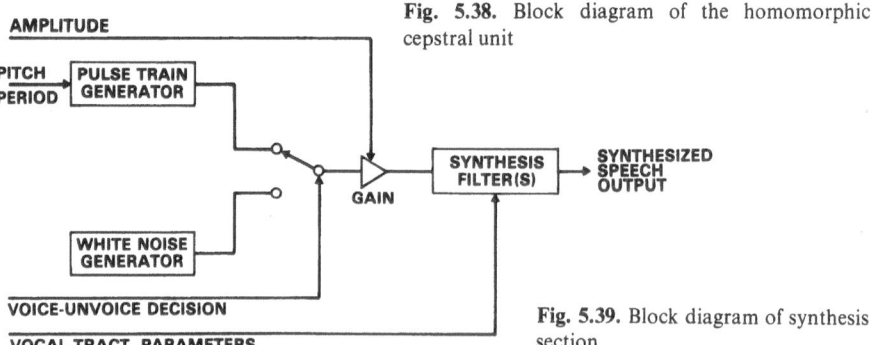

Fig. 5.39. Block diagram of synthesis section

These ten coefficients plus pitch and voicing information are the data which are encoded and transmitted to the receiver where the speech is reconstructed.

The calculation of each partial correlation coefficient involves addition, multiplication, and division. Each of this operations can be accomplished in a straightforward manner by applying the appropriate controls to the arithmetic chip. The divide operation requires a table lookup in addition to the arithmetic operations. The table lookup can be realized either with a conventional read-only memory (ROM) or with a CCD serial ROM.

The present major limitations to an all-digital approach are size, power dissipation, and cost. It is possible, however, that very large scale integration (VLSI) technology may impact some of these limitations in the future. Figures 5.40 and 41 illustrate the use of the CCD adaptive filter as an analysis filter to generate the prediction coefficients $W_1 \ldots W_N$. The voice input is bandpass filtered and applied to the adaptive filter. The filter input is also used as the

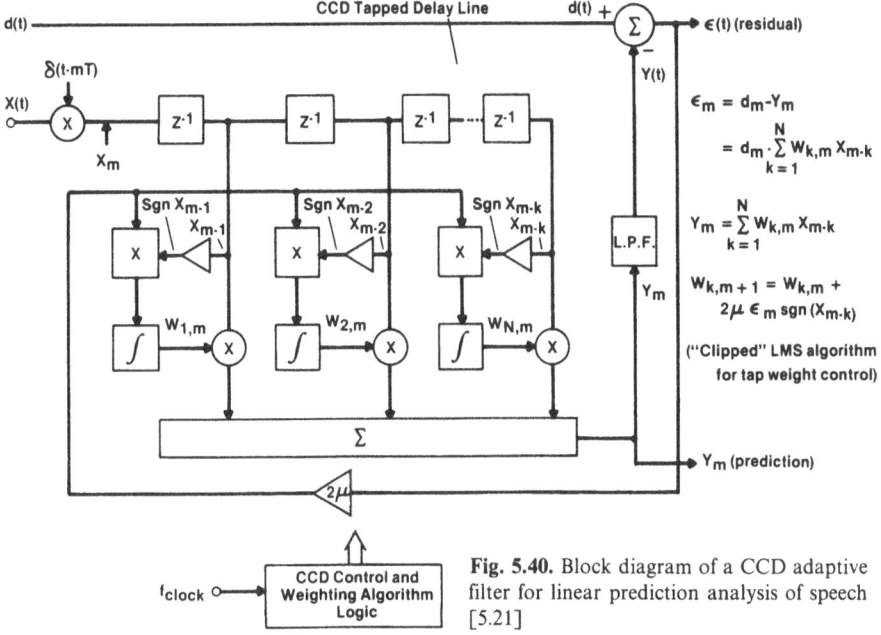

Fig. 5.40. Block diagram of a CCD adaptive filter for linear prediction analysis of speech [5.21]

$$\epsilon_m = d_m - Y_m$$

$$= d_m \cdot \sum_{k=1}^{N} W_{k,m} X_{m-k}$$

$$Y_m = \sum_{k=1}^{N} W_{k,m} X_{m-k}$$

$$W_{k,m+1} = W_{k,m} + 2\mu \, \epsilon_m \, \text{sgn} (X_{m-k})$$

("Clipped" LMS algorithm for tap weight control)

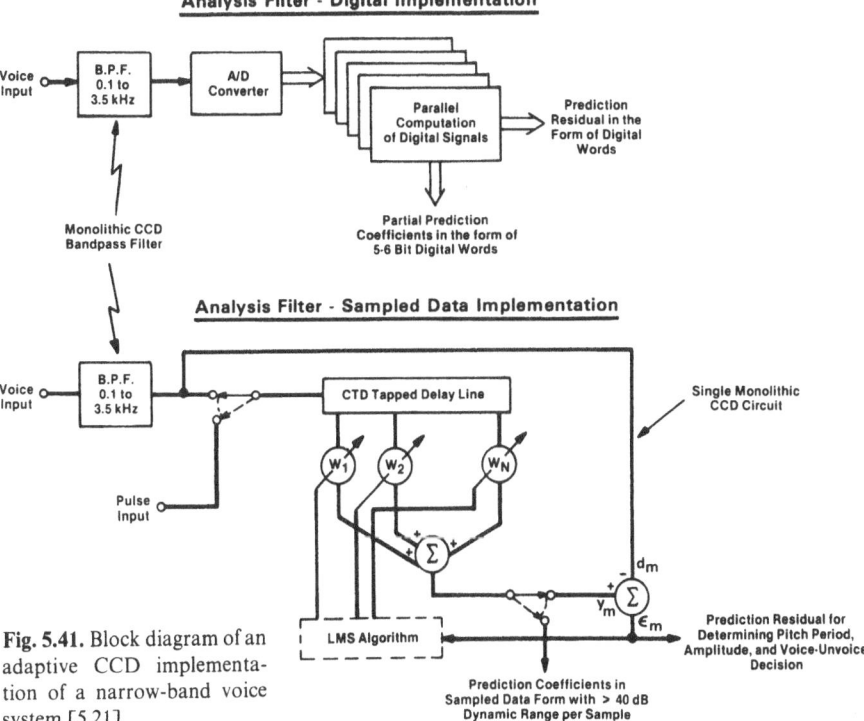

Fig. 5.41. Block diagram of an adaptive CCD implementation of a narrow-band voice system [5.21]

primary or desired signal to generate an error called the prediction residual. This error signal may then be used to extract the pitch period, amplitude, and voice/unvoice decision. The speech sample frame may be nominally 20 milliseconds in length and the error must converge to its minimum value within this time. Near the end of the time period, a unit pulse is inserted into the filter. The filter output is the impulse response of the filter. The converged weights are the predictive coefficients representing the state of the vocal tract. These coefficients can then be transformed to partial correlation coefficients for later transmission. The clipped data LMS algorithm has been computer simulated with the following constraints:

1) $W_k \leqq 0.98$.

2) An increase of the unit circle (Z-domain) by 10% with scaling of the prediction coefficients.

3) 10 prediction coefficients of bit levels: 8, 8, 8, 8, 7, 7, 7, 6, 5, 5, and

4) Data rate of 3600 bits/s.

The prediction coefficients generated with the clipped data LMS algorithm and the above constraints were used to synthesize speech. Test sentences were employed and the playback of the reconstructed speech indicated good speech reproduction and quality, although the latter is a subjective parameter. Thus, a CCD adaptive filter with 10 weights operating at an 8 kHz sample rate is a candidate for this particular application [5.21].

The use of the prediction residual of the analysis CCD adaptive filter as the prewhitened input to a pitch extractor is an important application. The prediction residual signal is the voiced signal with the vocal tract contributions removed. Ideally, this signal corresponds to the source of voiced sounds caused by the vibrating vocal chords at the glottis. Linear prediction implementations usually have access to the prediction residual. However, those speech processing techniques which are based upon the short-time spectrum analysis (e.g., the channel vocoder) do not yield a whitened version of the voiced signal as a by-product of their processing. In these systems, a CCD adaptive filter simply used as a prewhitener before pitch extraction by, for example, autocorrelation is extremely useful.

A major limitation to the use of LPC in practical acoustical environments is the degradation of synthesized voice quality due to the ambient acoustic noise background in the present of the speaker, This speech quality degradation is due primarily to the vocal-tract analyzer's inability to compute accurate prediction coefficients of the "uncontaminated" speech. The residual or "whitened" speech signal of the analyzer is likewise contaminated, which causes errors in pitch extraction for LPC systems that synthesize pitch from the residual signal. Several techniques can be applied to improve LPC performance in a noisy acoustical environment. These methods may be classified into two broad areas: 1) acoustical and/or electrical interference mitigation prior to computation of the vocal-tract parameters, and 2) algorithms, such as *Saber* [5.67], which attempt to separate the interference in the process of vocal tract computation. Electrical techniques, which mitigate interference prior to ana-

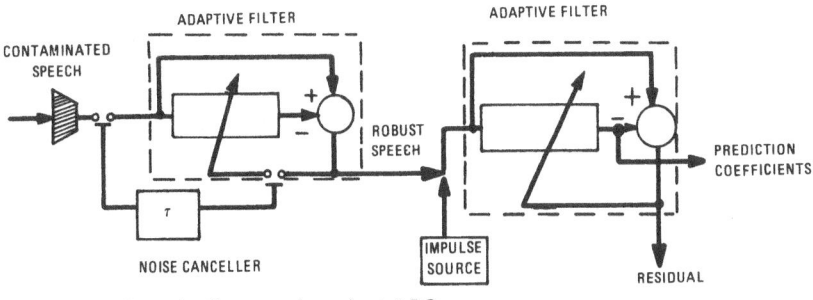

Fig. 5.42. Schematic diagram of a robust LPC system front end [5.21]

lysis, generally require two microphones. One microphone receives speech contaminated with ambient noise while the second microphone "receives" only the ambient noise. These signals are subsequently processed with an adaptive filter of the type described here for noise cancellation. In most practical environments it is difficult to arrange two strategically located microphones. A single microphone scheme can be implemented which will provide interference cancellation. This method assumes that a "push-to-talk" microphone (i.e., similar to popular CB sets) is used and a time lapse of 200 to 300 ms exists from activation of the microphone to the commencement of speech. A CCD adaptive filter, constructed of 6 to 20 taps, converges to a least-square estimate of the ambient noise within this 200 to 300 ms time lapse. Subsequently, the weights of the filter are fixed during the speech transmission and the filter serves to remove the background interference. One method for holding the weight values might be through the use of a digital memory coupled to MDACs at the taps of the CCD. This technique assumes that the noise environment is stationary in the period of speech transmission. In tests performed with the *Saber* algorithm, this requirement of noise stationarity did not limit the performance achieved in noise reduction. Figure 5.42 illustrates a robust LPC system block diagram which includes a noise canceller prior to the analyzer.

5.3.4 Radar

Two recent examples of potential applications for CCDs in radar are described here. One, which illustrates an interface with SAW devices, is a processor for pulse-Doppler radar. In this example, the CCD is required to sample the video signal, store the outputs from many radar returns, rearrange these outputs so the signals within each range window can be read out separately, and read out the signals at rates compatible with a SAW chirp-z transformer. The other application consists of a time-base expander for a monopulse tracking radar in which the CCD must sample the wide-band video at high speeds, integrate the *samples from* several radar returns, and read out the signals at slow rates in

N STAGES CORRESPONDING TO N RANGE GATES

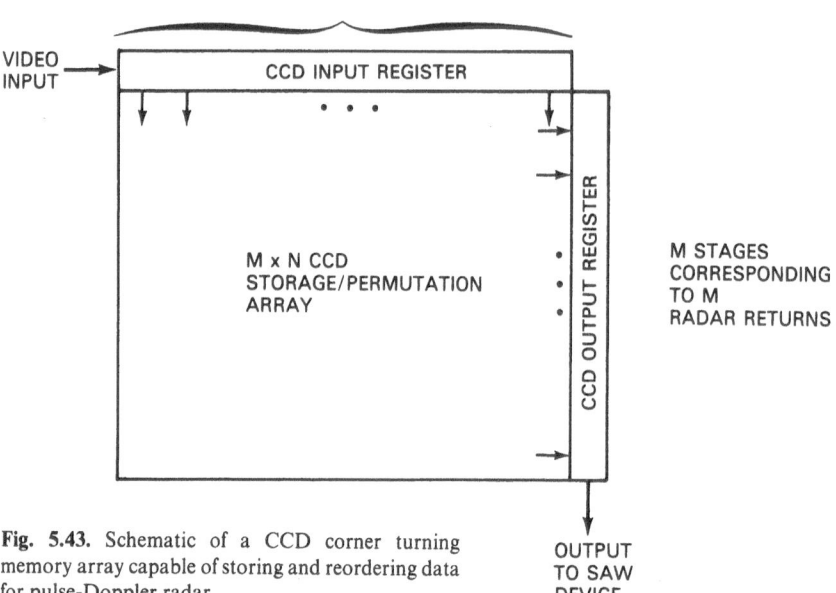

Fig. 5.43. Schematic of a CCD corner turning memory array capable of storing and reordering data for pulse-Doppler radar

order to ease the requirements on A/D converters and subsequent processors. Both of these applications would benefit by the further development of high-speed CCDs.

Pulse-Doppler Radar

In a pulse-Doppler radar system [5.68] an rf pulse is periodically transmitted. The return pulse is used to obtain two pieces of information about the target: distance to the target or range, and velocity of the target or range rate. The entire range of interest is divided into segments and the return signal is time gated into each range window sequentially. Range is thus determined simply by the delay time of the return pulse. If there is a target within a particular range, successive radar returns contain Doppler information which can be recovered from the bipolar video using a bank of narrow-band analog Doppler filters for each range gate. This obviously requires considerable hardware for a system with high Doppler resolution and a large number of range bins. The hardware can be simplified by using a digital FFT instead of the analog filter bank at each range gate, but further simplification is possible through the use of a CCD-SAW analog chirp-*z* transform device.

One implementation of the CCD part of the device is the corner turning memory array [5.69, 70] shown in Fig. 5.43. The radar video signal is clocked into the input CCD register at a rate suitable for the radar video bandwidth until the *N* stages corresponding to the *N* range gates of interest are filled. The charge is then clocked one step in parallel out of the input register into the storage array. This process is repeated for each radar return until M returns have been stored in the $M \times N$ array. To read out, the contents of the storage array are clocked to the right one stage, loading the output CCD register with the M returns from the first range gate. This signal is used to obtain the Doppler information by clocking it out rapidly into a SAW chirp-z transform device like that shown in Fig. 5.26a, with the post-multiplying chirp omitted. This is repeated for each range gate until all N signals have been transformed.

An alternative configuration, which is functionally equivalent to Fig. 5.43, uses N separate serial in/serial out CCD shift registers of length M instead of the two-dimensional array. The N shift registers are loaded by a serial in/parallel out input register [5.71, 72], or an MOS switching network [5.73], in the manner just described. When all the registers are filled with data, they are clocked out one at a time through a switching network into the SAW chirp-z transform device. The two-dimensional array in Fig. 5.43 is more difficult to implement than the N parallel shift registers because it requires more complicated charge clocking. The charge storage capability and thus the dynamic range may also be reduced because the storage sites are all bounded in two dimensions. However, the $M \times N$ array is basically a low-speed device; for high-speed operation only the input and output registers need to be clocked fast. For the pulse-Doppler radar application it is particularly desirable to have a high-speed output register to provide the most effective interface with the SAW device. The parallel shift register configuration requires high-speed outputs on all N channels. Both of these implementations have been operated successfully with dynamic range of 40 dB [5.72, 73]. Efforts are underway to improve the dynamic range to at least 50 dB in order to achieve acceptable performance even in clutter dominated environments.

The information obtained by this Doppler processor is the magnitude of the target velocity for each range gate. In order to determine whether the target is advancing or receding, a complex system is necessary, i.e., the SAW post-multiplying chirp must be included and the base-band processing must be done with I and Q channels. For continuous operation it is necessary to use a minimum of two identical CCD modules so that radar returns can be loaded into one array while the second is being emptied.

The maximum number of resolvable Doppler frequencies is equal to M, the number of radar returns corresponding to hits on target. For scanning radars, M is a function of radar parameters such as pulse repetition frequency, beamwidth, and scanning rate. This number is typically less than 100. The number of range bins, N, can vary widely depending on the radar design. For search radars more than 1000 range bins may be required. This can be achieved in principle by providing a serial output to the CCD input register of Fig. 5.43 so that several corner turning array chips can be cascaded. The CCD-SAW chirp-z processor is readily adaptable to differing radar bandwidths and pulse repetition frequencies, so that it could be very useful for jittered prf systems or multimode radars.

APERTURE OF FOUR-PORT ANTENNA FEED

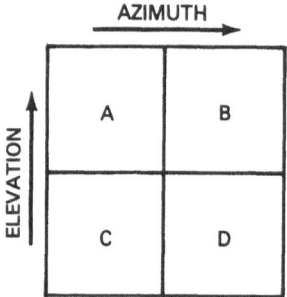

SUM BEAM   (A + B + C + D)   → RANGE

DIFFERENCE BEAM ([A + B] − [C + D]) → ELEVATION

DIFFERENCE BEAM ([A + C] − [B + D]) → AZIMUTH

Fig. 5.44. Aperture of a four-port antenna feed used to explain the operation of a monopulse tracking radar

The use of the CCD corner turning storage module allows use of a single fast SAW CZT device to process the Doppler information from a large number of range gates. This formidable processing capability using versatile, compact hardware should prove to be a very important technique for pulse-Doppler radar.

Monopulse Tracking Radar

In an amplitude comparison monopulse radar system [5.74] short rf pulses are transmitted with repetition frequency of approximately 1 kHz and the return signal is used to obtain three pieces of information about the target; range, azimuth, and elevation. The operation can be explained using the diagrams in Fig. 5.44. The squares A, B, C, and D represent antenna feeds which receive the target return from a very narrow transmitted beam. Three signals are formed. $(A + C) − (B + D)$, called a difference signal, gives azimuth information. If the quantity is positive the target is to the left of center and if it is negative the target is to the right of center. This information is used by servos to move the antenna and track the target by keeping the return at the antenna center. Similarly $(A + B) − (C + D)$ gives elevation information. Range is determined by delay time of the return pulse using the sum signal $(A + B + C + D)$. A single range window corresponding to the position of the target being tracked is processed. The precise timing of the range gate is continuously updated to keep the target within the processed range.

For a high-resolution monopulse system [5.75] the transmitted pulse is very short ($\sim$3 ns) so that 45 cm target resolution is possible. Within a typical range window of 0.25–1 µs individual target features, such as airplane wing tanks, fuselage, and tail section, can be resolved separately. This high resolution requires very high video bandwidth, with sampling rates around 800 MHz. The computer needed for this system cannot process the data that fast but if a very high-speed CCD time- base expander could sample the data in real time and clock it out much slower, processing could take place at a more reasonable speed during the time period between radar returns. In addition to high-speed input buffering, the potential capability of the CCD to perform accurate analog processing of the signal is of major importance. An example of desired processing in this application is the capability for averaging the data in each resolution cell over ten pulse repetition periods. This integration process increases the signal-to-noise ratio and provides an additional order of magnitude reduction in the required processing speed of the computer. High speed CCDs could have a significant impact in this type of application, but only if the linearity and dynamic range can be made competitive with digital processors. After integration a dynamic range equivalent to eight bits (48 dB) is required by the system.

The high-speed CCD input buffer for this application should be capable of storing 500–1000 samples of the input signals at sampling rates of about 800 MHz. After the data are clocked in for the entire range window of interest, the high-speed clock is stopped. A low-speed (approximately 1 MHz) clock is then started which reads out the data to the integrator. The integration is preferably done with a separate CCD from the input buffer because the high-speed clocking is likely to heat the chip enough to reduce the integration time by thermal carrier generation. Also, separating the high-speed CCD and the relatively low-speed integrator CCD allows them to be optimized independently. There are several possible integrator configurations. A recirculating loop has the advantage that fixed pattern noise due primarily to leakage current nonuniformities can be cancelled by putting a unity gain inverter in the loop and recirculating the data twice for each data repetition interval.

The CCD just described represents some highly ambitious specifications which for a single channel device can only be speculative at this time, but the possibility of multiplexing several slower and smaller capacity devices to achieve the required performance does exist either by combining several chips [5.76] or by fabricating a single multichannel chip [5.66, 78]. The CCD in this application not only samples and slows down the signal, but performs an analog integration on the data. This type of processing would allow operation at the full desired sampling rates without resorting to difficult, costly, high capacity digital processors.

5.4 Conclusions and Projections

The split-electrode CCD transversal filter has been shown to be a highly useful device. It has been demonstrated that this device can have a dramatic impact on the cost of systems which use matched filters and Fourier transforms. Split-electrode transversal filters, adaptive transversal filters and DCCL are potentially useful for advanced secure voice processing systems.

The feasibility of the CCD adaptive transversal filter concept has been demonstrated. It is expected that filters having tens of taps will be fabricated on a single chip. It is expected that these devices will be highly useful as noise cancellers in communication systems and in medical electronics.

A framework for performing digital logic with the charge-coupled concept has been investigated. Compared with analog devices, digital devices have the virtues of choice of accuracy, insensitivity to temperature variations, and general application. Compared with other digital technologies, using the same lithographic rules, DCCL has the advantages of lower-power requirements and higher functional density. This technology could be highly useful in applications which require a high degree of accuracy and which have stringent power and size constraints. An example of such an application would be secure voice communications.

Sonar systems, especially lightweight or expendable systems, will incorparate CCDs into beam formers and signal processing. Even very large beam formers may involve CCDs for cost and size reduction.

It has been shown that charge packets can be transferred at rates of hundreds of MHz in charge-coupled device structures. However, input techniques, output techniques, and clocking techniques at these speeds are not straightforward. Nevertheless, it is expected that these problems will be overcome to the point where high-speed CCDs will be highly useful in radar and electronic-warfare systems. Also it is expected that the CCD-SAW combination will provide more signal processing capability than could be achieved with either technology alone.

Research efforts are presently underway which may lead to significant advances in CCDs for signal processing. In the infrared area, MIS technologies for both InSb and HgCdTe are being developed with the objective of being able to fabricate CCDs for focal plane processing. Current interface state densities are around $1 \times 10^{11} \, \text{cm}^{-2} \text{eV}^{-1}$. Less than an order of magnitude improvements is necessary to make relatively long low-loss CCDs feasible in these materials.

Advanced infrared focal planes, both scanned and staring, will undoubtedly make extensive use of CCD processing. Systems for imager, search and track, seeker, and surveillance applications will provide better performance at acceptable cost by utilizing CCD techniques for the focal plane and following processing.

References

5.1   D. B. Scott, S. G. Chamberlain: IEEE J. SC-**12**, 45–50 (1977)
5.2   Y. T. Chan, B. T. French, P. E. Green: "Extremely High Speed CCD Analog Delay Line", in 1975 Intern. Conf. Applic. CCDs Proc., pp. 389–398
5.3   Y. T. Chan: "A Sub-nanosecond CCD", in 1976 NASA Conf. CCD Technol. Applic. Proc., pp. 89–94
5.4   M. F. Tompsett: IEEE Trans. ED-**22**, 305–309 (1975)
5.5   C. H. Sequin, A. M. Mohsen: IEEE J. SC-**10**, 81–92 (1975)

5.6   J.C.Fraser, D.H.Alexander, R.M.Finnila, S.C.Su: "An Extrinsic Silicon Charge-Coupled Device for Detecting Infrared Radiation", in IEEE 1974 IEDM Proc., pp. 442–445
5.7   S.P.Emmons, D.D.Buss: J. Appl. Phys. **45**, 5303 (1974)
5.8   S.P.Emmons, A.F.Tasch, J.M.Caywood, C.R.Hewes: "A Low-Noise Input with Reduced Sensitivity to Threshold Voltage", in IEEE 1974 IEDM Proc., pp. 233–234
5.9   D.F.Barbe: Proc. IEEE **63**, 38–67 (1975)
5.10 R.W.Brodersen, S.P.Emmons: IEEE J. SC-**11**, 147–155 (1976)
5.11 K.R.Hense, T.W.Collins: IEEE J. SC-**11**, 197–202 (1976)
5.12 C.H.Sequin: "Antialiasing Inputs for Charge-Coupled Devices", in IEEE 1976 IEDM Proc., pp. 31–34
5.13 H.Wallinga: "A Comparison of CCD Analog Input Circuit Characteristics"; in 1974 Intern. Conf. Technol. Applic. CCDs Proc., pp. 13–21
5.14 D.D.Buss, D.R.Collins, W.H.Bailey, C.R.Reeves: IEEE J. SC-**8**, 138–146 (1973)
5.14a D.Feder: "X-Ray Lithography", in X-ray Optics, ed. by H.T.Queisser, Topics in Applied Physics, Vol. 22 (Springer, Berlin, Heidelberg, New York 1977)
5.15 D.D.Buss, C.R.Hewes, M.deWit: "Charge-Coupled Devices for Analogue Signal Processing", in IEEE 1976 Intern. Specialists Seminar Impact New Technol. Signal Processing Proc., pp. 17–26
5.16 R.W.Brodersen, C.R.Hewes, D.D.Buss: IEEE J. SC-**11**, 75–84 (1976)
5.17 L.R.Rabiner, R.W.Schafer, C.M.Rader: IEEE Trans. AU-**17**, 86–92 (1969)
5.18 H.J.Whitehouse, J.M.Speiser, R.W.Means: "High Speed Serial Access Linear Transform Implementations", Naval Underseas Center Technical Note 1026 (1973)
5.19 A.A.Ibrihim, G.J.Hupe: "Fully Integrated Voiceband Filters", in 1978 Intern. Conf. Appl. CCDs Proc., pp. 3A–15–3A–16
5.20 J.N.Gooding, T.E.Curtis, W.D.Pritchard, R.A.Rehman: "Programmable Transversal Filter Using CCD Components", in 1978 Intern. Conf. Applic. CCDs Proc., pp. 3B-23–3B-30
5.21 M.H.White, I.A.Mack, G.M.Borsuk, D.R.Lampe, F.J.Kub: "CCD Analog Adaptive Signal Processing", in 1978 Intern. Conf. Applic. CCDs Proc., pp. 3A-1–3A-14
5.22 M.H.White, I.A.Mack, F.J.Kub, D.R.Lampe, J.L.Fagan: "An Analog CCD Transversal Filter with Floating Clock Electrode Sensor and Variable Tap Gain", in 1976 Intern. Sol. St. Cir. Conf. Proc. pp. 194–195
5.23 J.L.McCreary, P.R.Gray: IEEE J. SC-**10**, 371–378 (1975)
5.24 D.R.Morgan, S.E.Craig: IEEE J. ASSP-**24**, 494–500 (1976)
5.25 C.S.Miller, T.A.Zimmerman: "The Application of CCDs to Digital Signal Processing", in 1975 Intern. Commun. Conf. Proc., pp. 20–24
5.26 T.A.Zimmerman, R.A.Allen, R.W.Jacobs: IEEE J. SC-**12**, 473–485 (1977)
5.27 D.F.Barbe, W.D.Baker: Microelectron. J. **7**, 36–45 (1975)
5.28 C.S.Miller, C.F.Motley, R.A.Allen: "Digital Charge-Coupled Device Technology and Digital Filter Applications", in 1977 EASCON Proc., pp. 30-1A to 30-1H
5.29 B.Gold, C.M.Rader: Digital Processing of Signals (McGraw-Gill, New York 1969)
5.30 R.C.Singleton: IEEE Trans. AU-**17**, 93–103 (1969)
5.31 L.J.M.Esser: Electron. Lett. **8**, 620–621 (1972)
5.32 L.J.M.Esser, M.G.Collet, J.G.van Santen: "The Peristaltic Charge-Coupled Device", in IEEE 1973 IEDM Proc., pp. 17–20
5.33 L.J.M.Esser: "Peristaltic Charge-Coupled Devices: What is Special About the Peristaltic Mechanism", in Solid State Imaging, ed. by P.G.Jespers, F.van de Wiele, M.H.White (Noordhoff, Leyden 1976) pp. 343–425
5.34 G.M.Borsuk, M.H.White, N.Bluzer, D.R.Lampe: "Design and Analysis of New High Speed Peristaltic CCDs" in 1978 Intern. Conf. Applic. CCDs Proc., pp. 4-85–4-94
5.35 N.Bluzer, H.C.Lin: "High Speed Bipolar CCD Input Structure", in IEEE 1978 IEDM Proc., pp. 624–627
5.36 I.Deyhimy, J.S.Harris, R.C.Eden: "GaAs CCD with High Transfer Efficiency", in 1978 Intern. Conf. Applic. CCDs Proc., pp. 2-79–2-84
5.37 R.L.Van Tuyl: "A Monolithic Integrated 4-GHz Amplifier", in IEEE 1978 Intern. Sol. St. Cir. Conf. Digest, pp. 72–73

5.38 R. C. Eden, B. M. Welch, R. Zucca: "Low Power GaAs Digital ICs Using Schottky Diode-FET Logic", in IEEE 1978 Intern. Sol. St. Cir. Conf. Digest, pp. 68–69

5.38a E. A. Ash: "Fundamentals of Signal Processing Devices", in *Acoustic Surface Waves*, ed by A. A. Oliner, Topics in Applied Physics, Vol. 24 (Springer, Berlin, Heidelberg, New York 1978)

5.39 R. C. Williamson: Proc. IEEE **64**, 702–710 (1976)

5.40 O. W. Otto, H. M. Gerard: "On Rayleigh Wave Reflection from Grooves at Oblique Incidence and an Empirical Model for Bulk Wave Scattering in RAC Devices", in IEEE 1977 Ultrasonics Symp. Proc., pp. 596–601

5.41 V. S. Dolat, R. C. Williamson: "BGO Reflective-Array Compressor (RAC) with 125 μs of Dispersion", in IEEE 1975 Ultrasonics Symp. Proc., pp. 390–394

5.42 O. W. Otto: "The Chirp Transform Signal Processor", in IEEE 1976 Ultrasonics Symp. Proc., pp. 365–370

5.43 M. A. Jack, G. F. Manes, P. M. Grant, C. Atzeni, L. Masotti, J. H. Collins: "Real Time Network Analyzers Based on SAW Chirp Transform Processors", in IEEE 1976 Ultrasonics Symp. Proc., pp. 376–381

5.44 R. M. Hays, C. S. Hartmann: Proc. IEEE **64**, 652–671 (1976)

5.45 J. B. G. Roberts: Electron. Lett. **12**, 509–510 (1976)

5.46 M. A. Jack, P. M. Grant: Electron. Lett. **13**, 65–66 (1977)

5.47 J. C. Carson: "Infrared Mosaic Technology", SPIE Proc., Vol. 62 (1975) p. 3

5.48 E. S. Kohn: IEEE J. SC-**11**, 139–146 (1976)

5.49 A. J. Steckl, R. D. Nelson, B. T. French, R. A. Gudmundsen, D. Schechter: Proc. IEEE **63**, 67–74 (1975)

5.50 W. Steffe, L. Walsh, C. K. Kim: "A High Performance 190 × 244 CCD Area Image Array", in 1975 Intern. Conf. Applic. CCDs Proc., pp. 101–108

5.51 R. D. Thom, R. E. Eck, J. D. Phillips, J. B. Scorso: "InSb CCDs and Other MIS Devices for Infrared Applications", in 1975 Intern. Conf. Applic. CCDs Proc., pp. 31–42

5.52 S. P. Emmons, A. F. Tasch, J. M. Caywood, C. R. Hewes: "A Low-Noise Input with Reduced Sensitivity to Threshold Voltage", in IEEE 1974 IEDM Proc., pp. 233–234

5.53 S. P. Emmons, T. F. Cheek, J. T. Hall, P. W. Van Atta, R. Bakerak: "A CCD Multiplexer with Forty AC Coupled Inputs", in 1975 Intern. Conf. Applic. CCDs Proc., pp. 43–52

5.54 Hughes Aircraft: "Low Cost Arrays for Detection in Infrared"; Interim Tech. Rpt. on Contract F33615-75-C-1175, Air Force Avionics Laboratory (1976)

5.55 J. Hunt, H. Sadowski: "Diverse Electronic Imaging Applications for CCD Image Sensors", in 1975 Intern. Conf. Applic. CCDs Proc., pp. 181–188

5.56 D. M. Erb, K. Nummedal: "Buried Channel Charge-Coupled Devices for Infrared Applications", in 1973 Intern. Conf. Applic. CCDs Proc., pp. 157–168

5.57 W. Grant, R. Balcerak, P. Van Atta, J. T. Hall: "Integrated CCD Bipolar Structure for Focal Plane Processing of IR Signals'.' in 1975 Intern. Conf. Applic. CCDs Proc., pp. 53–58

5.58 A. F. Milton, M. Hess: "Series-Parallel Scan IR CCD Focal Plane Array Concept", 1975 Intern. Conf. Applic. CCDs Proc., pp. 71–84

5.58a A. F. Milton: "Charge Transfer Devices for Infrared Imaging", in *Optical and Infrared Detectors*, ed. by R. J. Keyes, Topics in Applied Physics, Vol. 19 (Springer, Berlin, Heidelberg, New York 1977)

5.59 O. Ohtsuki, H. Sei, K. Tanikawa, Y. Miyamoto: "CCD with Meander Channel", in 1976 Intern. Conf. Technol. Applic. CCDs Proc., pp. 38–43

5.60 J. Mattern, D. R. Lampe: IEEE J. SC-**11**, 88–92 (1976)

5.61 S. P. Buchanan, R. R. Clark: "A CCD Analog Memory System for Slow-Scan Conversion of Standard TV Video", in 1976 Intern. Conf. Technol. Applic. CCDs Proc., pp. 364–370

5.62 M. H. White, D. R. Lampe: "Charge-Coupled Device Analog Signal Processing", in 1975 Intern. Conf. Applic. CCDs Proc., pp. 189–198

5.63 D. R. Lampe, M. H. White, J. L. Fagan, J. H. Mims: "An Electrically Reprogrammable LSI Analog Transversal Filter", in IEEE 1974 Intern. Sol. St. Cir. Conf. Digest, pp. 156–157

5.63a R. P. Porter: "Acoustic Probing of Space-Time Scales in the Ocean", in *Ocean Acoustics*, ed. by J. A. DeSanto, Topics in Current Physics, Vol. 8 (Springer, Berlin, Heidelberg, New York 1979)

5.64 G.S.Kang: "Linear Predictive Narrowband Voice Digitizer", in 1974 EASCON Proc. pp. 51–58

5.65 F.Itakura, S.Saito: "On the Optimum Quantization of Feature Parameters in PARCOR Speech Synthesizer", in IEEE 1972 Speech Conf. Proc., pp. 434–437

5.66 D.F.Barbe: "The Use of CCDs in Voice Processing Systems", in IEEE 1977 Intern. Symp. Circuits Systems Proc., pp. 410–417

5.67 S.Ball: "Application of SABER Method for Improved Spectral Analysis of Noisy Speech", Tech. Report UUCS-77-107, 1977

5.68 M.I.Skolnik (ed.): *Radar Handbook* (McGraw-Hill, New York 1970)

5.69 J.D.Collins, W.A.Sciarretta, D.D.MacFall, M.B.Schulz: "Signal Processing with CCD and SAW Technologies", in IEEE 1976 Ultrasonics Symp. Proc., pp. 441-450

5.70 W.L.Eversole, D.J.Mayer, R.J.Kansey: "A CCD Two Dimensional Transform", in 1978 Conf. Applic. CCDs Proc., pp. 3B-31–3B-40

5.71 J.B.G.Roberts, R.Eames, D.V.McCaughan, R.F.Simons: IEEE J. SC-**11**, 100–104 (1976)

5.72 D.V.McCaughan, A.J.W.Turner, J.M.Keen, R.Eames, J.B.G.Roberts: "Developments in Radar Doppler Processing", in 1978 Intern. Conf. Applic. CCDs Proc., pp. 3B-53–3B-62

5.73 W.H.Bailey, R.J.Kansy, R.A.Kempf, R.C.Bennett, J.L.Owens: "A Complementary CCD/SAW Radar Signal Processor", in 1978 Intern. Conf. Applic. CCDs Proc., pp. 3B-41–3B-52

5.74 J.H.Dunn, D.D.Howard, K.B.Pendleton: "Tracking Radar", in *Radar Handbook*, ed. by M.I.Skolnik (McGraw-Hill, New York 1970)

5.75 D.D.Howard: "High Range-Resolution Monopulse Tracking Radar and Applications", in 1974 EASCON Proc., pp. 86–91

5.76 T.E.Linnenbrink, M.J.Monahan, J.L.Rea: "A CCD-Based Transient Data Recorder", in 1975 Intern. Conf. Applic. CCDs Proc., pp. 443–453

5.77 J.W.Balch, C.F.McConaghy: "A CCD Integrated Circuit for Transient Recorders", in 1976 NASA Conf. CCD Technol. Applic. Proc., pp. 115–117

5.78 Y.T.Chan, B.T.French, A.J.Hughes, W.N.Lin, M.J.McNutt, W.E.Meyer: "A 1024-Cell CTD Shift Register Capable of Digital Operation at 50 MHz", in 1978 Intern. Conf. Applic. CCDs Proc., pp. 4-71–4-83

Additional References:

F.Kub, D.Boyle, M.Evey, E.Naviasky: "Monolithic Peristaltic CCD/ECL Technology", in Custom Integrated Circuits Conference, Rochester, NY (May 23–24, 1979)

T.E.Linnenbrink, D.A.Gradl, G.J.DeWitte, D.S.Metzger, E.K.Hodson, D.R.Thayer, J.W.Balch, C.F.McConaghy: "A One Gigasample Per Second Transient Data Recorder", IEEE Trans. NS-**26**, 4443–4449 (1979)

M.D.Pocha, J.W.Balch, C.F.McConaghy: "Two New CCD Archetectures for High Speed Transient Recording", in IEEE 1979 Intern. Sol. St. Cir. Conf. Digest, pp. 14–16

G.Borsuk, F.Kub, N.Bluzer, D.Lampe, M.Evey, M.White: "High Speed Silicon CCDs", in Device Research Conference, Boulder, Colorado (June 25–27, 1979)

6. Radiation Effects in Silicon Charge-Coupled Devices

J. M. Killiany

With 23 Figures

Charge-coupled devices (CCDs) have potentially extensive applications in optical imaging, signal processing, and serial memories. They have small size, low power consumption, and high reliability. Such characteristics make CCDs attractive for certain space, military, and nuclear industrial applications provided the devices can satisfy the radiation hardness requirements. The CCD radiation tolerance required will of course be dependent on the radiation environment and the amount of shielding available. An unshielded device in certain earth orbits can receive a dose in excess of 10^6 rads (Si) per year from electrons and protons trapped in the Van Allen radiation belts. Nuclear industrial environments may subject a device to neutron fluences in excess of 10^{11} neutrons per square centimeters and gamma ray doses greater than 10^4 rads (Si). It is relatively easy to shield devices from electrons and protons. However, gamma ray and neutron protection is more difficult to provide.

The radiation-produced damage in a semiconductor device depends on its relative sensitivity to ionization and bulk displacement effects. Surface-controlled devices (MOS) tend to be limited by the ionization produced, since ionization leads to charge buildup in insulator layers and to an increase in surface state density at the insulator-semiconductor interface. The characteristics of bulk-effect devices are usually degraded by displacement damage, since this damage can significantly decrease carrier concentration, carrier mobility, and carrier lifetime. CCDs are sensitive to both surface and bulk radiation damage effects. They are also very susceptible to transient radiation-induced loss of stored information.

This chapter has been organized in the following manner. First, the basic radiation-induced degradation mechanisms in silicon devices are reviewed. Then the surface, bulk, and transient radiation effects in CCDs are discussed. Radiation hardening techniques are presented in Sect. 6.5 and liquid nitrogen temperature radiation effects are described in Sect. 6.6. Finally, the current status of understanding of radiation effects on CCDs is summarized.

6.1 Radiation Effects in Semiconductor Devices

The basic effects of radiation on semiconductor devices have been described by *Gregory* and *Gwyn* [6.1]. The salient features of that review will be briefly *outlined here* as a background for the discussion of expected radiation effects in

CCDs. Additional background material may be obtained from the books by *Larin* [6.2] and *Chaffin* [6.3].

High-energy radiation deposits energy in semiconductor materials via two mechanisms: atomic collisions and electronic ionization. The atomic collisions displace atoms from their normal positions in the semiconductor crystal to other locations in the lattice. Ionization, of course, is the removal of orbital electrons from the semiconductor atoms to form ions and free electrons. The relative importance of these two mechamisms depends on both the type of radiation and the nature of the device. Electrons, protons, and gamma ray deposit most of their energy via the ionization process, while fast neutrons deposit up to 50 % of their energy in atomic displacement damage. MOS devices are sensitive to surface damage resulting from cumulative ionizing radiation exposure, while the characteristics of bulk-effect devices such as bipolar transistors, are primarily degraded by displacement damage. CCDs are sensitive to both surface and bulk displacement damage effects. They are also very susceptible to transient ionizing radiation-induced loss of stored information.

Numerous studies have shown that ionizing radiation causes failure of MOSFET (metal oxide semiconductor field effect transistor) devices due to two mechanisms: 1) trapped charge buildup in the silicon dioxide layer, and 2) an increase in the density of trapping states at the silicon-silicon dioxide interface. Ionizing radiation generates electron-hole pairs in the silicon dioxide. All the electrons are rapidly swept out of the oxide by the applied field but a fraction of the holes are permanently trapped producing a negative threshold voltage shift. The size of the threshold voltage shift varies with the magnitude and polarity of the applied gate bias during irradiation. Positive gate to substrate bias results in a larger threshold voltage shift since the holes are trapped near the silicon surface where they will exert maximum influence on the semiconductor. With negative gate bias during irradiation, the holes tend to be trapped near the metal gate where they have much smaller effect on the surface potential and mobile charge in the silicon. Interface state generation during the irradiation of MOS structure is also enhanced for positive gate to substrate bias. These states, located at the plane separating the semiconductor from the insulator, can interact with mobile carriers in the silicon reducing the surface mobility which in turn causes a decrease in MOSFET transconductance. Charge trapped in interface states can also shift the threshold voltage in MOS transistors. In *p*-channel devices the trapped charge is positive thereby adding to the negative threshold voltage shift due to the positive oxide space charge buildup. For *n*-channel structures the occupied traps are negatively charged and may partially compensate the negative threshold voltage shift. In some cases, the interface state charge can dominate and cause a net positive threshold voltage shift. The absolute magnitude of the oxide charge buildup and the interface state density increase is greatly dependent on the details of the MOS fabrication process.

Ionization effects are usually characterized in terms of units of absorbed dose (rads). One rad corresponds to the absorption of 100 ergs per gram of material. This unit depends on neither the type of radiation nor the material in which the energy is absorbed. However, it is important to specify the material when this term is used. The absorbed dose in terms of rads (Si) is usually employed when reporting MOS ionizing radiation effects. Ionization effects per rad are identical, independent of what radiation caused the ionization.

In addition to these permanent degradation mechanisms, ionizing radiation produces electron-hole pairs in the silicon substrate during irradiation. Minority carriers generated in the device depletion region or within a diffusion length of these regions can produce photocurrents at the device terminal. Bursts of radiation can cause large transient currents which can result in logic upset and even device burnout due to heating effects. Dose rates may exceed 10^{11} rads (Si) per second for a 100 ns pulse. The large currents associated with a transient radiation pulse can trigger a p–n–p–n structure in a high-power dissipation state where it remains even after the primary photocurrents have decayed. This effect is called "latch-up."

The type of displacement damage produced in silicon is a function of the radiation species. Gamma rays and electrons of a few MeV produce simple defects which are best described as a distribution of point defects. The present model for bulk damage in silicon resulting from neutron bombardment is the formation of superclusters, each containing a number of highly damaged regions (subclusters). Protons produce a mixture of point and cluster defects. Hence, the equivalence of the different types of radiation in causing displacement damage cannot be given by a single quantity as was done in the case of ionization effects unless the defect dominating the damage is the same for both types of radiation.

Displacement damage in silicon leads to significant decreases in carrier concentration, carrier mobility, and minority carrier lifetime. The dominant failure mechanisms in most neutron-irradiated bulk semiconductor devices are a reduction in minority carrier lifetime ($> 10^{11}$ n/cm^2) and decrease in carrier concentration ($> 10^{13}$ n/cm^2). Mobility degradation in silicon does not become severe until neutron fluence exceeds 10^{15} n/cm^2.

The damage in a silicon semiconductor device soon after exposure to a pulse of neutrons can be more severe than would be predicted based on passive long-term exposure data. The ratio of the damage, at any time after exposure, to the final stable damage is called the annealing factor (AF). The annealing factor 1 ms after a neutron burst can be as large as 7. For a room temperature irradiation an annealing factor near 1 is reached a few seconds after exposure. The recovery time constant increases as the device temperature is lowered.

6.2 Surface Damage Effects

Since CCDs are MOS devices, ionizing radiation causes a buildup of positive charge in the gate oxide and an increase in trapping states at the silicon-silicon

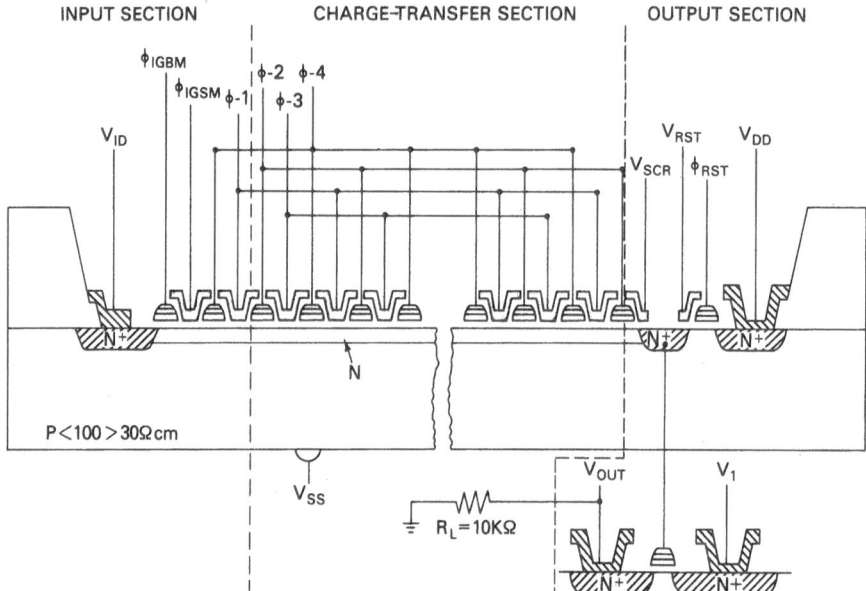

Fig. 6.1. Cross section of a four-phase *n*-buried-channel CCD shift register

dioxide interface. The resulting negative flat-band voltage shift causes a change in the CCD operating bias while the interface state density increase reduces the charge-transfer efficiency in surface channel devices and increases the dark current density in both surface and buried-channel structures. Consequently, nonhardened devices are unsuitable for most room temperature applications requiring a dose tolerance greater than 5×10^4 rads (Si). The details of these radiation-induced failure mechanisms are presented in the following paragraphs.

6.2.1 Positive Charge Buildup in the Oxide

For a given oxide technology, the smallest flat-band voltage shifts are expected for *n*-buried channel and *p*-surface channel devices because the gate voltage is negative with respect to the channel potential. The effects of positive space-charge buildup in the oxide on CCD operation will be demonstrated using a simple shift register of the type shown in Fig. 6.1. The CCD shift register is separated into the following three parts for the purpose of discussing radiation effects.

a) Input structure, consisting of the input diode, input gates, and first two clock gates.

b) Charge-transfer section, consisting of the clocked electrodes.

c) Output detector-amplifier, consisting of the output gate, reset gates, output diode, reset MOSFET, and source follower MOSFET.

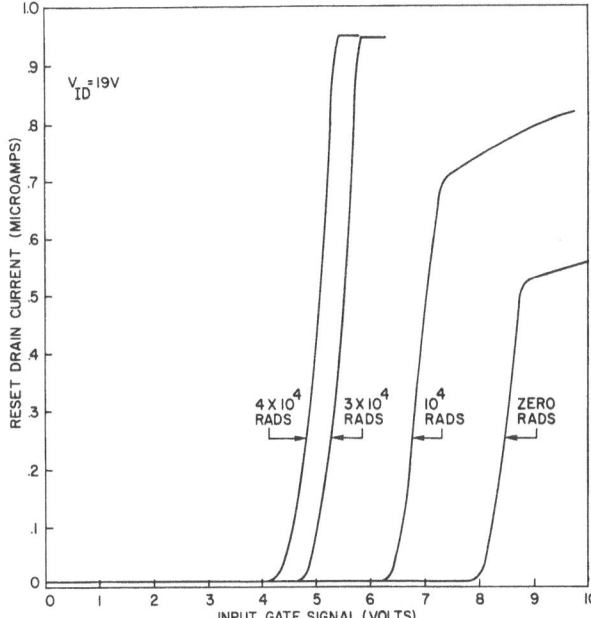

Fig. 6.2.
CCD transfer curves for the dynamic injection input as a function of radiation dose

a) *Input Structure:* The input structure of the shift register is the most radiation-sensitive section of the device unless a threshold-insensitive input is employed. Input techniques, such as dynamic injection, operate by leaking charge into the CCD potential well over the barrier formed by the input gates ϕ_{IGBM} and ϕ_{IGSM}. According to *Sequin* and *Mohsen* [6.4], the quantity of charge injected is given by

$$Q_s = \frac{\beta}{2}(V_{\text{IGID}} - V_T)^2\, T,$$

where V_{IGID} is the difference in the input gate and the input diode voltages and V_T is the input gate threshold voltage. β is a constant dependent on the device geometry and T is the time available for filling the potential well. An input of this type is extremely sensitive to small changes in input gate to input diode bias and typically tolerates only 10^3 rads (Si) before requiring adjustment. An example of the radiation-induced shift in the CCD transfer curve for the dynamic injection input is shown in Fig. 6.2. A negative threshold voltage shift will eventually cause input saturation for n-surface and n-buried channel devices and input cutoff for p-surface and p-buried channel structures.

Killiany and *Baker* [6.5] have shown that threshold shifts up to -5 V may be accommodated by the input structure when the potential equilibrium input, a threshold-insensitive technique, was employed. The manner in which the potential equilibrium input technique operates is illustrated in Fig. 6.3. A potential well is established under the phase 4 and 1 electrodes ($\phi-4$ and

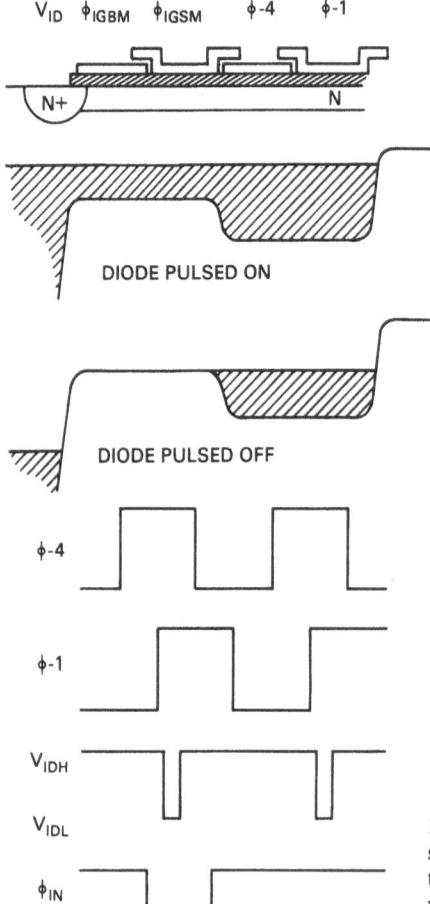

Fig. 6.3. Cross section of CCD input structure showing potential energy profile in the silicon for the potential equilibration input technique along with the gate and diode voltages

$\phi - 1$). The input diode (V_{ID}) is pulsed to cause charge to flow over the barrier established under the input gates (ϕ_{IGSM} and ϕ_{IGBM}). The charge is prevented from spilling down the CCD channel by the barrier under the phase 2 gate ($\phi - 2$). While both transfer gates ($\phi - 4$ and $\phi - 1$) are still high, the diode voltage is returned to the resting level (V_{IDL}) and the excess charge is drained from under $\phi - 4$ and $\phi - 1$. The amount of charge retained in the CCD well is proportional to the difference in the channel potential under the $\phi - 4 - \phi - 1$ electrodes and the input gates. Hence, to first order, the amount of signal charge injected in the CCD will be threshold voltage insensitive for flat-band voltage shifts of equal magnitude under the input and first transfer gates.

The CCD transfer curves generated while using the potential equilibration input technique are shown in Fig. 6.4. The post-irradiation transfer curves coincide with the preirradiation curve in the threshold-insensitive region for threshold voltage shifts up to -3.6 V. The maximum threshold voltage shift

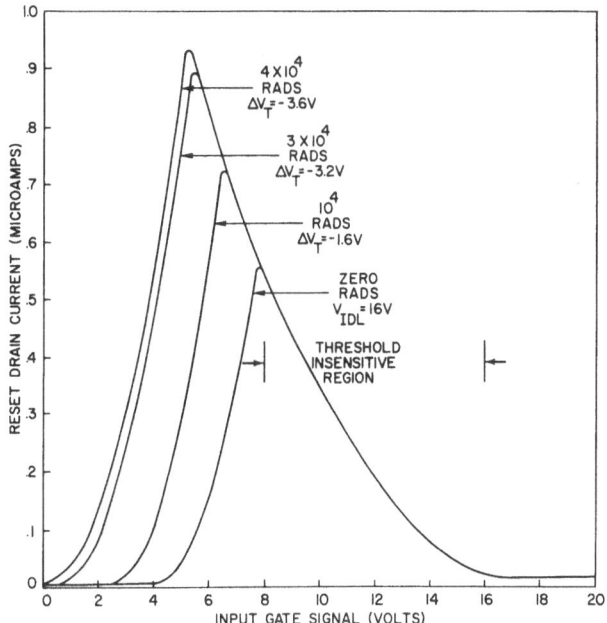

Fig. 6.4.
CCD transfer curves for the potential equilibration input as a function of radiation dose

which could be accommodated in this particular device while maintaining a reasonable signal injection capacity was $-5.0\,\mathrm{V}$. The potential equilibration input has a limited flat-band voltage shift accommodation because the channel potentials ϕ are a function of the flat-band voltage while the input diode voltage levels V obviously are not. In the case of the n-buried channel device shown in Fig. 6.3, a negative flat-band voltage shift reduces the barrier under phase 2 ($\phi-2$) which eventually allows charge to spill down the CCD channel. The magnitude of the threshold voltage shift accomodation could of course be increased by adjusting the input diode low level; this would, however, also reduce the signal handling capacity of the input. While the operation of the potential equilibration input was illustrated using a buried channel device it should be noted that this technique is also applicable to surface channel structures.

b) *Charge-Transfer Section:* The charge-transfer process in a properly designed charge-coupled device is fairly insensitive to uniform transfer gate flat-band voltage shifts. However, gate to gate nonuniformities may distort the potential profile in the CCD channel with increased trapping of the signal charge as a result. *Barbe* et al. [6.6] have shown that unequal flat-band voltage shifts on adjacent gates can cause a severe loss in CCD signal handling capacity. The difference in flat-band voltage shifts between the aluminum and polysilicon gates in a stepped oxide structure is shown in Fig. 6.5. After 3×10^5 rads the unequal flat-band voltage shifts reduced the signal handling capacity to 20% of the pre-irradiation value. A cross section of the two-phase stepped oxide shift

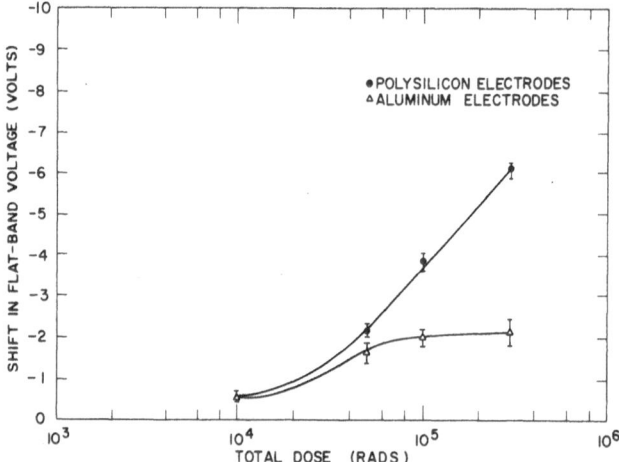

Fig. 6.5. Unequal flat-band voltage shifts for polysilicon and aluminium gate electrodes as a function of dose

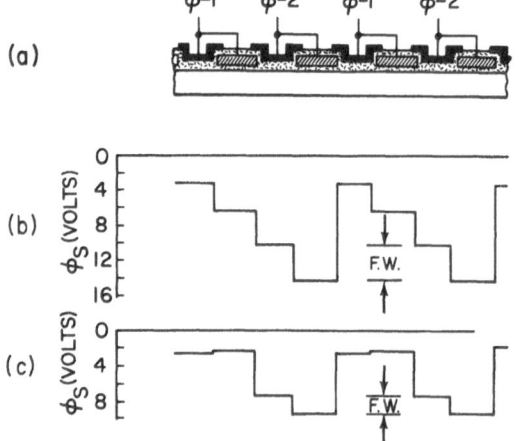

Fig. 6.6a–c. Distortion of the channel potential profile in a two-phase surface channel CCD caused by unequal flat-band voltage shifts for adjacent gate electrodes

register along with the pre- and post-irradiation channel potential is shown in Fig. 6.6.

A negative flat-band voltage shift will cause the potential energy profile in the CCD channel to change in both surface and buried channel structures. Eventually the channel will be driven out of depletion in p-surface and n-buried channel devices. *Killiany* et al. [6.7] have shown that the increase in reset drain voltage V_{DD} required to operate an n-buried channel after irradiation is approximately equal to the threshold voltage shift (see Fig. 6.7). A few volts of flat-band voltage shift accommodation can be obtained in an n-buried channel structure simply by applying a reverse bias V_{DD} to the reset drain, several volts in excess of the preirradiation value required fo deplete the channel. The flat-

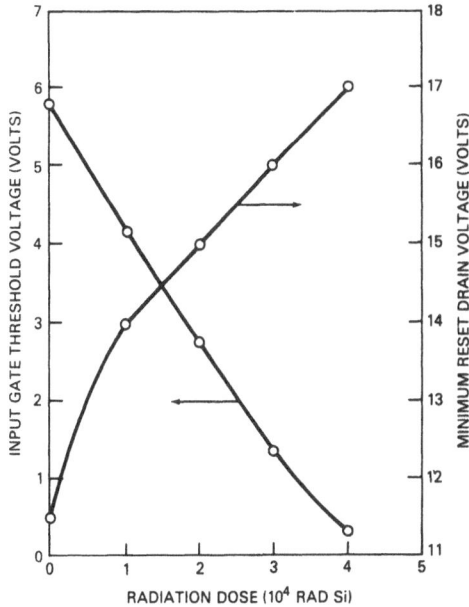

Fig. 6.7. Minimum reset voltage V_{DD} required to deplete the channel in an n-buried channel CCD as a function of dose compared to input gate threshold voltage

band voltage shift tolerance of a p-surface channel device can be increased by either applying a positive substrate bias or a negative clock offset voltage.

c) *Output Section:* The CCD output is the least radiation sensitive part of the device. The reduced radiation sensitivity of the CCD output can be attributed to the following:

1) The reset MOSFET is used as a switch to reset the gate of the source follower transistor to the reference voltage V_{DD}. Hence, the reset pulse levels can be set to values which will alow pre- and post-irradiation operation.

2) The output MOSFET threshold voltage shift is smaller than the shift on the other CCD gates due to the reduced electric field strength in that oxide. Electric field values in the transfer gate and source follower gate oxides of a typical p-surface channel device are 0.9×10^6 V/cm and 0.25×10^6 V/cm, respectively.

3) The source follower output is ac coupled. Hence, small shifts in the dc operating point of the output MOSFET are of little consequence.

4) Since the output transistor is operated as a source follower, the output gain A_V will be relatively insensitive to changes in the transconductance g_m for values of $g_m R_L > 1$

$$A_V \simeq \frac{g_m R_L}{1 + g_m R_L},$$

where R_L is the source follower load resistor.

6.2.2 Interface State Density Increase

The irradiation-produced increase in trapping state density at the silicon-silicon dioxide interface reduces the charge-transfer efficiency in surface channel devices and increases the surface component of the dark current density in both surface and buried channel CCD structures. The transfer inefficiency ε_s attributable to interface state trapping is directly proportional to the interface state density. An empirical expression for the transfer inefficiency was given by *Carnes* and *Kosonocky* [6.8], namely

$$\varepsilon_s = \frac{1}{n_s + n_{s,0}} kTN_{ss} \ln\left(1 + \frac{2f_c}{k_1 n_{s,0}}\right),$$

where $n_{s,0}$ is the number of bias charge carriers per unit area, n_s is the number of signal carriers per unit area, N_{ss} is the interface state density in $cm^{-2} (eV)^{-1}$, f_c is the clock frequency, and k_1 is a constant parameter that depends on the trapping cross section ($k_1 \simeq 10^{-2} cm^2/s$).

Most of the interface state trapping in surface channel devices operated with an adequate bias charge takes place at the edges of the gate electrodes. Interface states in those regions are not "covered" by the bias charge because the edges of the potential wells are not vertical. This phenomena is called "edge effect".

The increase in charge transfer inefficiency due to interface state trapping in surface channel devices renders them unsuitable for most applications after 10^5 rads (Si). Increasing the amount of bias charge to a value greater than that required for preirradiation operation does not appreciably improve the transfer efficiency (edge effect). Surface state trapping effects do not occur in a properly operated buried channel device since the charge packet is transferred in the bulk silicon rather than at the silicon-silicon dioxide interface. *Hartsell* [6.9] has compared the transfer inefficiency as a function of dose for surface and buried channel devices (see Fig. 6.8). The small degradation of the transfer efficiency observed in the buried channel device after large doses is probably due to gamma-induced bulk trapping effects.

Interface states also act as generation centers for the surface component of the dark current. The dark current density due to interface state N_{ss} generation is given by

$$J_{DS} = \frac{qn_i}{2}(\pi kT\sigma_n v_{th})N_{ss},$$

where n_i is the intrinsic carrier concentration, q is the electronic charge, k is Boltzmann's constant, T is absolute temperature, σ_n is the capture cross section for electrons, and v_{th} is the average thermal velocity for electrons.

Killiany et al. [6.7] observed the linear relation between the increase in interface state density and the increase in dark current density with dose illustrated in Fig. 6.9. The interface state density was determined using the

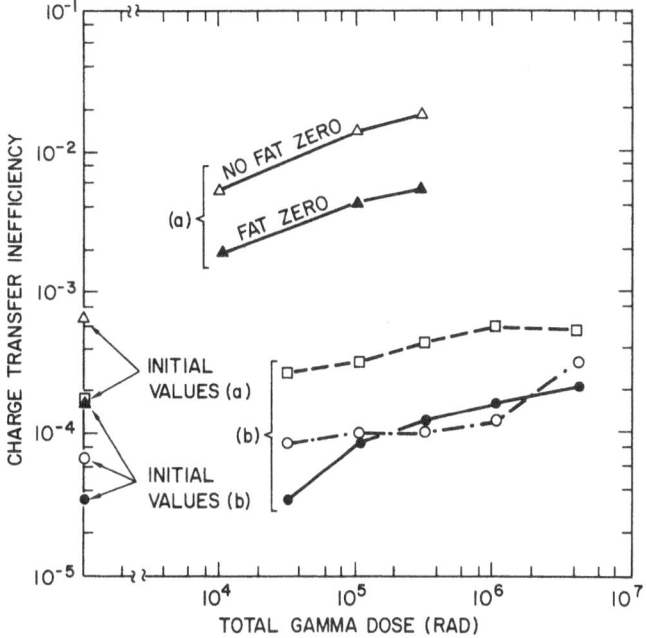

Fig. 6.8. Charge transfer inefficiency vs gamma dose (a) for surface channel CCDs, (b) for buried channel CCDs

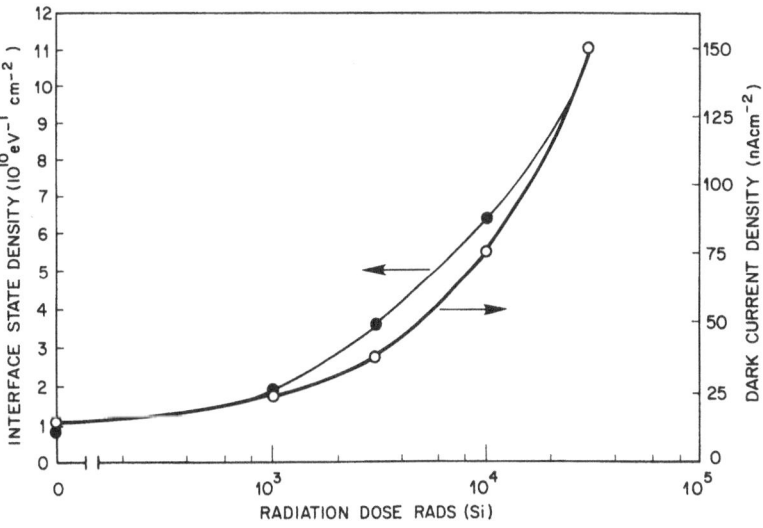

Fig. 6.9. Increase in dark current density and interface state density in a surface channel CCD as a function of dose

periodic pulse technique for surface channel devices developed by *Carnes* and *Kosonocky* [6.8].

The thermally generated charge in a CCD limits the length of time a signal charge can be stored in the device. *Hartsell* [6.9] measured dark current densities of $1000 \, \text{nA/cm}^2$ after 10^6 rads (Si) in devices having a preirradiation value of $2 \, \text{nA/cm}^2$. Increases of this magnitude usually prohibit application of unhardened devices after doses of 10^5 rads (Si). *Hartsell* et al. [6.10] also measured the CCD noise in a buried channel device after irradiation. They observed an increase by an amount equivalent to the shot noise on the increased dark current.

Ionizing radiation does not increase the number of localized regions of high dark current density (i.e., dark current spikes).

6.3 Transient Ionization Effects

Charge-coupled devices are extremely sensitive to transient radiation-induced loss of stored information since they are very sensitive photosensors. *Williams* and *Nelson* [6.11], using basic theoretical considerations, determined that the CCD potential wells should become saturated with radiation-induced charge at doses of the order of 1 rad (Si). The ionizing radiation pulse creates electron-hole pairs in the bulk silicon. Those minority carriers generated in a CCD potential well or within a diffusion length L of the potential well will be collected. Hence, the ionizing radiation dose γ at which the central potential wells in a CCD area array become saturated is given by

$$\gamma = \frac{N_{\text{FW}}}{g_\gamma (L + W) A_{\text{bit}}},$$

where N_{FW} is the full well capacity expressed in number of charge carriers, W is the width of the depletion region, g_γ is the electron-hole pair (ehp) generation rate constant for ionizing radiation, and A_{bit} is the surface area of one CCD bit (photosite).

For a buried channel device with $N_{\text{FW}} = 4.5 \times 10^5$ electrons, $L = 50 \, \mu\text{m}$, $A_{\text{bit}} = 16.25 \, \mu\text{m}^2$ and $g_\gamma = 4.3 \times 10^{13}$ electron-hole pairs $\text{cm}^{-3} \text{rad (Si)}^{-1}$. the wells are predicted to saturate after a dose of 0.32 rad (Si).

The saturation dose for photosites near the perimeter of an area array and in a linear imager is lower since photocharge generated within a diffusion length of the array edge may contribute to the well filling. Surface channel devices have a higher saturation level since in general they have a larger charge handling capacity per unit gate area. The CCD saturation dose can be increased for both surface and buried channel devices by reducing the thickness of the silicon substrate to a value which is smaller than the thickness $(W + L)$ of the original photosite collection volume. A reduction of the effective substrate thickness to $10 \, \mu\text{m}$ is currently possible by means of chemical etching or by the use of an epitaxial structure.

RECOVERY TIME vs CLOCK FREQUENCY

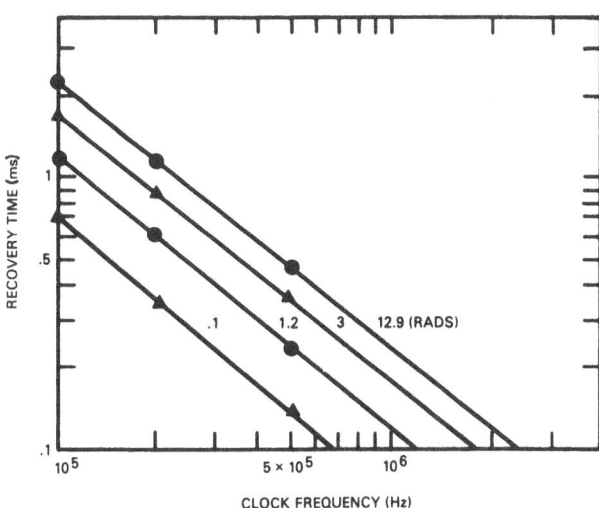

Fig. 6.10.
CCD recovery time as a function of clock frequency and pulse dose

Random single-bit loss of stored information has been observed in CCDs bombarded by heavy ions. *May* and *Woods* [6.12] recently determined that the "soft errors" (i.e., correctable) observed in CCD memories can be attributed to the alpha particles emitted by radioactive elements in the device packaging material. A 5 MeV alpha particle has only a 25 μm path length in silicon but creates 1.4×10^6 electron-hole pairs. This is sufficient charge to saturate a CCD potential well. *Pickel* and *Blandford* [6.13] recently developed a model for predicting the cosmic ray-induced bit error rate in dynamic MOS RAMs (random access memories). This model can also be used to determine the soft error rate for CCD memories in satellite systems.

Device burnout can be prevented during a transient radiation pulse by simply current limiting the power supply leads to the CCD diodes. *Hartsell* [6.9] successfully used this technique for dose rates up to 10^{11} rads (Si) per s (100 ns pulse width). Latch-up had neither been expected in CCDs nor been observed during high dose rate tests due to the absence of a parasitic *n-p-n-p* structure in the CCD.

In some applications the amount of time required to recover normal device operation after a radiation pulse is important. *Shedd* and *Buchanan* [6.14] observed the removal of excess charge generated in a CCD at a rate which is a function of the CCD clock frequency f_c for a given dose (see Fig. 6.10). Additional analysis indicates a recovery time dependence on CCD well capacity N_{FW}, since the quantity of charge carriers transported out of a CCD during a clock period is limited by the well capacity. Hence, the CCD recovery time T_R after a pulse of ionizing radiation can be approximated by

$$T_R \simeq \frac{g_\gamma V_{coll}}{N_{FW} f_c},$$

where V_{coll} is the collection volume associated with one CCD photosite. The data in Fig. 6.10 also show that the recovery time does not increase linearly with radiation pulse dose. This effect can probably be attributed to electron-hole pair recombination and direct carrier removal through the output diode for large overload doses.

The continuous background radiation level that results in CCD well saturation depends on the time a given depletion region exists in the CCD. For a device operating as an imager (integration mode), the wells fill with charge primarily during the integration time while in the continuous mode of operation the time a potential well remains in the device is given by the product of the device length in (bits) and the shift register clock period.

In addition to filling the potential wells, the charge generated by the continuous background radiation can become the dominant noise source in a CCD. *Autio* and *Bafico* [6.15] have shown that the noise in a radiation-generated CCD charge packet is larger than the shot noise which accompanies either a thermally or optically generated charge packet of the same magnitude. For a quantum efficiency of unity, one electron-hole pair is generated in silicon by each optical photon. Therefore, the rms fluctuation $N_n(B)$ in the number of charge carriers in a packet containing an average number N carriers is

$$N_n(B) = \sqrt{N}.$$

However, a high-energy ionizing particle creates an average number $\bar{N}_\gamma$ of electron-hole pairs in a CCD photosite collection volume. Therefore, the average number N_E of ionizing particles passing through the photoside collection volume is given by

$$N_E = N/\bar{N}_\gamma.$$

The rms fluctuation in the number of ionizing events per collection volume is given by $\sqrt{N_E}$. Hence, the number of noise carriers $N_n(\gamma)$ associated with the N radiation generated carriers is

$$N_n(\gamma) = \sqrt{N_E}(\bar{N}_\gamma).$$

The ratio of the radiation-induced shot noise associated with a packet of N carriers to the rms fluctuations in the number of carriers in an equal size optically generated packet is called the shot noise multiplier G_S

$$G_s = \frac{N_n(\gamma)}{N_n(B)}$$

or

$$G_S = \sqrt{\bar{N}_\gamma}.$$

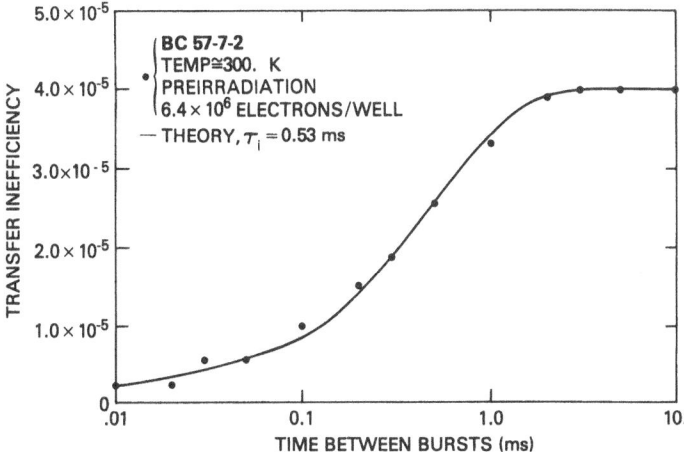

Fig. 6.11. Periodic pulse data for an unirradiated *n*-buried channel CCD and the fit of the discrete bulk trap level theory (solid line) to the data. The bulk trap emission time constant obtained from this fit was 0.53 ms at 300 K

According to *Autio* and *Bafico* [6.15] each Compton electron created in silicon by cobalt-60 gamma rays (1.25 MeV) generates 900 electron-hole pairs in a CCD collection volume which is 10 µm thick. Hence, $G_S = \sqrt{900} = 30$ for their particular test device.

6.4 Displacement Damage Effects

Fast neutrons create bulk trapping centers in silicon which cause a decrease in transfer efficiency primarily in buried channel devices and an increase in the bulk component of the dark current density in both surface and buried channel structures. *Mohsen* and *Tompsett* [6.16] related the transfer inefficiency ε in a buried channel CCD to the bulk trap density N_t for a single level of traps by

$$\varepsilon = \frac{V_{SIG}}{N_s} N_t e^{-T_t/\tau_i}(1 - e^{-T_E/\tau_i}),$$

where V_{SIG} is the volume occupied by the signal charge packet, N_s is the number of signal charges in volume V_{SIG}, T_t is the time available for charge transfer between adjacent electrodes, τ_i is the bulk trap emission time constant, and T_E is the time the bulk traps have to empty between signal charge packets.

The measurement of the transfer inefficiency as a function of the time the bulk traps are allowed to empty is known as the "periodic pulse" technique. The fit of the transfer inefficiency equation to the preirradiation periodic pulse data of *Saks* et al. [6.17] is shown in Fig. 6.11. A preirradiation bulk trap density of 5.6×10^{11} cm^{-3} is calculated from these data.

Table 6.1. Bulk trap energy levels and creation rates

Level	$(E_c - E_t)$ [eV]	$\Delta N_t / \Delta\phi$ [cm^{-1}]
N-1	0.14	1.1
N-2	0.23	~0.8
N-3	0.41	7.0

Hartsell [6.9] observed that increased trapping effects in neutron-irradiated *n*-buried channel CCDs are insignificant for fluences less than 10^{11} n/cm^2. *Saks* et al. [6.17] reported a linear transfer inefficiency increase in the 10^{11} to 10^{13} n/cm^2 fluence range (~ 15 MeV average) (see Fig. 6.12). After 10^{13} n/cm^2 (~ 15 MeV average) the transfer efficiency at 295 K was reduced to 0.992, making the devices unsuitable for most applications. From the work of *Smith* et al. [6.18], the degradation measured after 15 MeV neutron bombardment is expected to be 2.5 to 3 times greater than the value for a 1 MeV equivalent neutron fluence.

Neutron irradiation produces several trapping levels of unequal density in silicon. *Saks* [6.19] observed three distinct trap levels from 77 to 300 K by use of the periodic pulse technique. The energy level $(E_c - E_t)$ and the creation rate $\Delta N_t / \Delta\phi$ of the bulk traps are given in Table 6.1. *Walker* and *Sah* [6.20] reported similar trapping levels in other types of irradiated silicon devices.

Bulk traps are created during neutron bombardment in both surface and buried channel CCD structures. However, the volume occupied by a charge packet in a surface channel device is more than an order of magnitude smaller than the volume in a buried channel structure. Consequently, the transfer efficiency in a surface channel device is less sensitive to increases in bulk trap density since the charge packets interact with fewer bulk traps. The comparison of the change in transfer inefficiency with neutron fluence for surface and buried channel devices shown in Fig. 6.13 is taken from *Hartsell* [6.9].

While 50 % of the fast neutron energy is deposited in silicon via displacement damage, the fraction for gamma rays is smaller. *Curtis* and *Srour* [6.21] determined that the bulk trap creation rate for a 1 MeV photon per cm^2 fluence is only 1×10^{-3} cm^{-1}. Hence, bulk damage effects due to gamma irradiation are insignificant for doses less than 10^6 rads (Si) (i.e., 2×10^{15} 1 MeV gammas cm^{-2}). *Stein* and *Gereth* [6.22] measured carrier removal rates in silicon for 1.7 MeV electrons in the 0.2 to 1.0 cm^{-1} range.

The neutron-generated bulk trap level located near mid-gap, $E_c - E_t = 0.41$ eV, acts as a bulk generation center for dark current in both surface and buried channel devices. *Saks* et al. [6.17] observed an approximate linear increase in the dark current density in the 10^{11}–10^{13} n/cm^2 (15 MeV average) range for an *n*-buried channel structure. The data in Fig. 6.14 also show that the surface component of the dark current did not increase with neutron fluence. Little change was expected since the total dose equivalent of

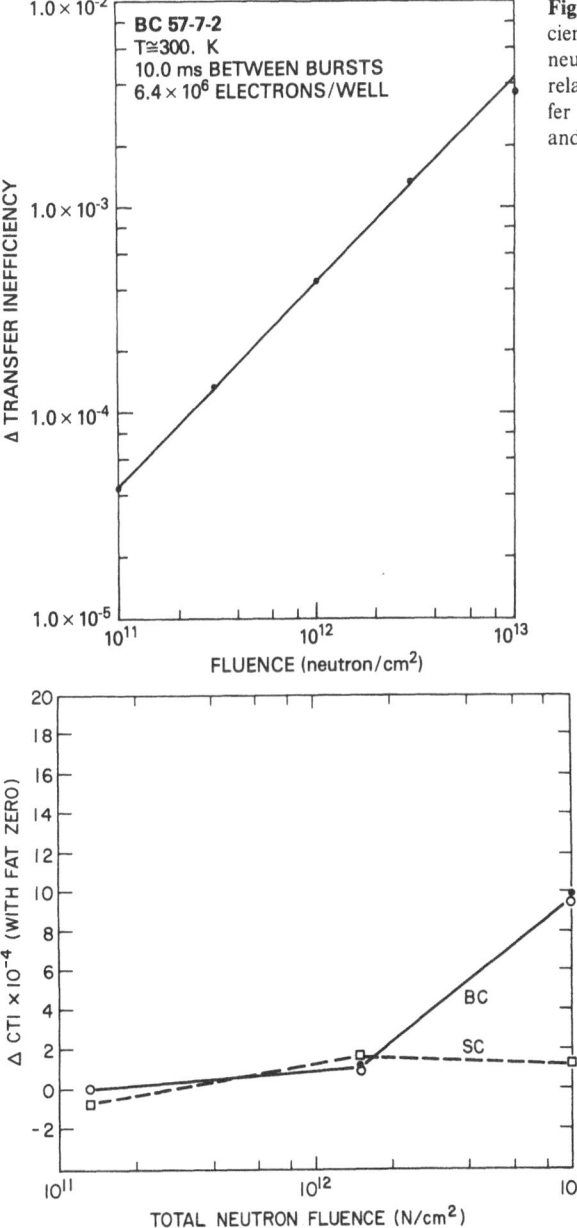

Fig. 6.12. Change in transfer inefficiency at 300 K as a function of neutron fluence, illustrating the linear relation between the increase in transfer inefficiency due to bulk trapping and the neutron fluence

Fig. 6.13. Change in charge transfer inefficiency (CTE) with neutron fluence (1 MeV equivalent) for typical surface and buried channel CCDs

10^{13} n/cm^2 is only approximately 10^4 rads (Si). A linear relation between the dark current density increase and the neutron fluence was predicted by *Williams* and *Nelson* [6.11] from a simple consideration of the irradiation dependence of the minority carrier lifetime τ in silicon,

$$\frac{1}{\tau} = \frac{1}{\tau_0} + K\phi,$$

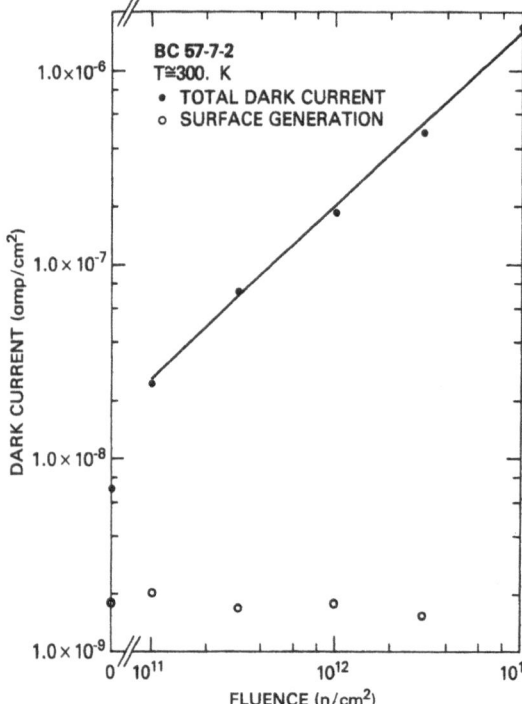

Fig. 6.14. Dark current density at 300 K as a function of neutron fluence, illustrating the linear relation between J_D and the neutron fluence

where τ_0 is the preirradiation minority carrier lifetime, K is the minority carrier lifetime degradation constant, and ϕ is the neutron fluence,

The bulk component of the dark current density is given by

$$J_D = \frac{qn_i d}{2\tau},$$

where q is the electronic charge, n_i is the intrinsic carrier concentration, and d is the depth of the depletion region.

Hence, the increase in the bulk component of the dark current density is related to the neutron fluency by

$$\Delta J_D = \frac{qn_i d}{2} K\phi.$$

Srour et al. [6.23] have compared the values of $(\Delta J_D / \Delta \phi)$ derived from data obtained by several workers using different neutron sources. A modified version of their comparison is shown in Table 6.2. The values are in reasonable agreement if differences in device structure, neutron energy, and gamma radiation associated with the neutron sources are taken into account. The value of the minority carrier lifetime degradation constant obtained from the data of

Table 6.2. Comparison of values for $\Delta J_D/\Delta\phi$

Ref.	$\Delta J_D/\Delta\phi$ [nA cm^{-2}/neutron]	Radiation source	Comments
Hartsell [6.9]	$\sim 9.6 \times 10^{-11}$	Fast-burst reactor	Buried channel
	$\sim 5.6 \times 10^{-11}$	Fast-burst reactor	Surface channel
Saks et al. [6.17]	1.7×10^{-10}	Cyclotron (0–30 MeV neutrons, 15 MeV peak)	Buried channel
Chang and *Aubuchon* [6.26]	$\sim 2 \times 10^{-11}$	Fast-burst reactor	Buried and surface channel
Srour et al. [6.23]	$\sim 4 \times 10^{-11}$	TRIGA reactor	Buried channel

Saks et al. [6.17] with $d = 2\,\mu m$ is 2.3×10^{-6} cm^2 s^{-1} which is in the range of the values for the minority carrier lifetime degradation constant measured by *Curtis* [6.24].

Little effort has been made to neutron harden buried channel CCDs. One possible technique would be to fabricate a thinner but more heavily doped buried channel so that the signal charge packet would interact with a smaller number of bulk traps, thereby reducing the transfer efficiency degradation. Another approach might be to fabricate the devices on Czochralski-grown wafers since the carrier removal rate in oxygen-rich silicon is smaller than the damage in float zone material [6.22].

Pulsed neutron irradiation data has been published for CCDs [6.34]. The transient annealing factor are larger than those observed in other semiconductor devices for the period from 10 to 1,000 s after the neutron burst.

6.5 CCD Surface Damage Hardening

6.5.1 Structural Optimization

Barbe et al. [6.6] and *Killiany* et al. [6.7] in the earliest works on total dose radiation effects in CCDs identified the major failure mechanisms peculiar to several different device structures since a standard CCD design had not then emerged. The objective was to determine the design which would optimize the total dose tolerance of the device. In particular it was found that:

1) A buried channel structure should be used. The charge transfer efficiency in a buried channel device, contrary to surface channel CCDs, is not degraded by an increase in the interface state trap density after irradiation.

2) An n-buried channel CCD structure is preferred. The flat-band voltage shift for a given oxide structure during irradiation is minimized when the gate electrodes are negative with respect to the channel potential.

3) The n-buried channel CCD output diode should be capable of being reverse biased to a voltage which will allow the channel to remain depleted after irradiation. The negative flat-band voltage shift in an n-buried channel structure causes the buried channel to be driven out of depletion. A few volts of flat-band shift can be automatically accommodated by biasing the output diode to a value several volts in excess of the preirradiation bias required to deplete the buried channel.

4) The design should use a planar channel insulator (no stepped oxide) and only one type of electrode material. This is necessary to eliminate differences in the flat-band voltage shift under adjacent electrodes, since such differences can reduce or eliminate potential barriers which can result in reduced well capacity and increased charge-transfer inefficiency.

5) The use of undoped polysilicon for interelectrode isolation should be avoided. *Killiany* et al. [6.7] observed channeling in the isolation regions of two different types of device after a total dose of 1 to 3×10^4 rads (Si) with resulting deterioration in device performance.

6) The input structure should be compatible with the operation of a threshold-insensitive input technique if an analog input is required.

6.5.2 Threshold Tracker

Another technique for allowing a CCD to operate in a radiation environment is to devise a circuit (threshold tracker) which would sense the radiation-induced threshold voltage shift and automatically change the voltage applied to the CCD gates to compensate for that shift. *Carnes* et al. [6.25] have designed and fabricated the threshold tracker circuit shown in Fig. 6.15. The output voltage V_{out} in stage 3 ideally should have a value

$$V_{out} = V_{G_0} + \Delta V_{TCCD},$$

where V_{G_0} is the preirradiation CCD gate voltage and ΔV_{TCCD} is the threshold voltage shift for the CCD gate. However, the actual CCD threshold voltage shift is approximated by the shift in the sensing transistor T_6. The oxide thickness t_{ox6} exactly matches that of the CCD gate it is designed to track. Also, for a threshold tracker located on the same chip as the CCD, both structures are subjected to identical processing. The function of the first and second stages of the tracker is to bias the gate of T_6 at a voltage which compensates for the threshold voltage of the load transistor T_5 and sets the preirradiation value of V_{out} equal to one-half the drain voltage ($V_{DD}/2$). V_{out} is applied to the CCD gates by means of a clamp circuit which can be located off the CCD chip. The clamp circuit allows the *dc* gate potentials to be adjusted for the radiation-induced

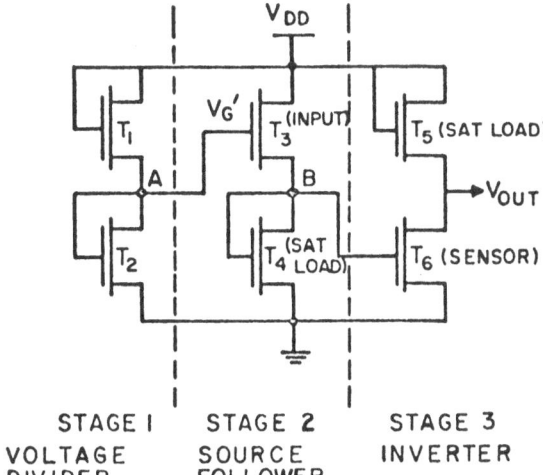

STAGE I   STAGE 2   STAGE 3
VOLTAGE   SOURCE   INVERTER   **Fig. 6.15.** Circuit diagram of the
DIVIDER   FOLLOWER   threshold tracking circuit

threshold voltage shift without changing the *ac* voltage swing of the clock waveforms.

This threshold tracker circuit was designed and successfully tested using discrete MOS components in a radiation environment. However, successful operation of the on-chip threshold tracker circuit in a radiation environment has not been obtained.

The primary drawback of the threshold tracker approach to total dose hardening is that the problem of the increased interface state density after irradiation is not addressed. Dark current density increases will severely limit the application of structures in which the threshold voltage shift has been accommodated. Also, the several threshold trackers and clamp circuits required to apply the compensated voltage to the various CCD gates introduce additional complexity to the circuit.

6.5.3 Hard Oxide Technology

Structural optimization alone is not sufficient to enable CCDs fabricated using standard gate oxide techniques to satisfy the total dose radiation requirements for most space and strategic applications. The approach taken by *Chang* and *Aubuchon* [6.26] in the development of a radiation-hard insulator for CCDs was to modify the process used in the fabrication of radiation-hard CMOS devices. An outline of the fabrication steps used for a radiation-hard *n*-buried channel process is shown in Fig. 6.16. Some of the process variations which have significant effect on the CCD hardness are the following:

1) A field oxide is grown and stripped to remove the damaged surface in the region where the gate oxide is to be grown.

MATERIAL: 2″, P-TYPE SILICON WAFERS
           < 100 > ORIENTATION
           30Ω CM RESISTIVITY

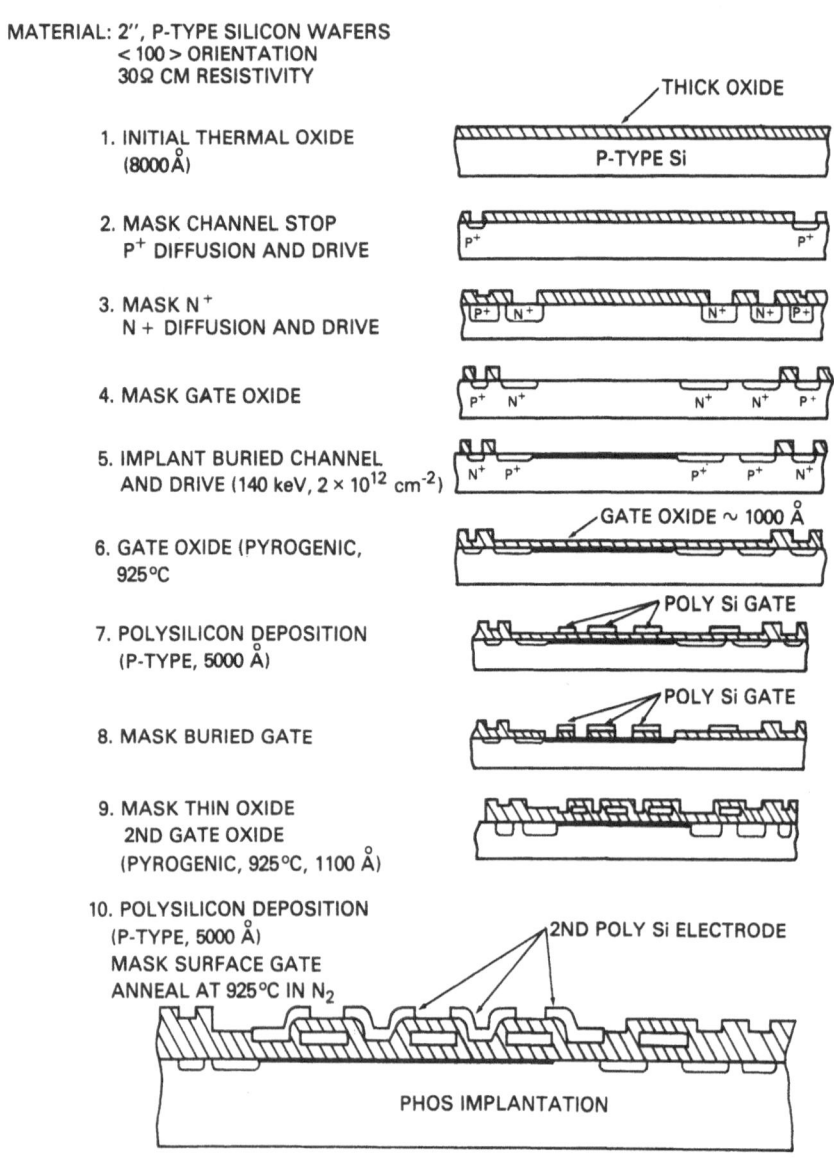

1. INITIAL THERMAL OXIDE
   (8000Å)

2. MASK CHANNEL STOP
   P+ DIFFUSION AND DRIVE

3. MASK N+
   N + DIFFUSION AND DRIVE

4. MASK GATE OXIDE

5. IMPLANT BURIED CHANNEL
   AND DRIVE (140 keV, 2 × 10¹² cm⁻²)

6. GATE OXIDE (PYROGENIC,
   925°C

7. POLYSILICON DEPOSITION
   (P-TYPE, 5000 Å)

8. MASK BURIED GATE

9. MASK THIN OXIDE
   2ND GATE OXIDE
   (PYROGENIC, 925°C, 1100 Å)

10. POLYSILICON DEPOSITION
    (P-TYPE, 5000 Å)
    MASK SURFACE GATE
    ANNEAL AT 925°C IN N₂

11. MASK CONTACT
12. DEPOSIT ALUMINUM
13. MASK ALUMINUM

14. ALLOY (475°C, FORMING GAS)
15. OVERGLASS
16. MASK PAD

Fig. 6.16. Outline of optimum radiation-hard *n*-buried channel CCD fabrication process

Table 6.3. Radiation-hard n-buried channel CCD parameters

	Preirradiation	1×10^6 rads (Si)
CTE	0.99999	0.9999
J_D	5 nA/cm^2	140 nA/cm^2
Well capacity	3×10^6 e$^-$	2.25×10^6 e$^-$
Threshold shift	–	-1.8 V

Table 6.4. Radiation-hard p-surface channel CCD parameters

	Preirradiation	1×10^6 rads (Si)
CTE	0.9997 20 % BC	0.992 50 % BC
J_D	20 nA/cm^2	100 nA/cm^2
Threshold shift	–	-2.0 V

2) The number of high-temperature processing steps required after the growth of the gate oxide is minimized by performing source-drain diffusions and bulk state gettering prior to the growth of the gate oxide. The choice of a 925 °C pyrogenic oxide was made for the same reason (i.e., faster growth rate).

3) The buried channel is implanted before the gate oxide is grown. Ion implantation through a radiation-hard oxide may result in reduced radiation hardness even after the implantation-produced flat-band voltage shift has been annealed.

4) The oxide thickness under all CCD gates is made approximately equal to minimize differences in flat-band voltage shift after irradiation. The oxide is made thin as is feasible in order to reduce flat-band voltage shift.

5) Eliminate the use of an electron gun for the aluminum metallization. *Mayo* et al. [6.27] have determined that an X-ray dose of 10^6 rads (Si) is absorbed during a typical e-beam process. This exposure seriously degrades the oxide radiation hardness for negative gate to channel bias during subsequent radiations.

Both surface and buried channel CCD shift registers have been fabricated using the hard oxide process. These devices can be operated after 10^6 rads (Si) with the preirradiation clock and bias voltages. The pre- and post-irradiation n-buried channel CCD parameters obtained by *Chang* [6.28] are shown in Table 6.3. The post-irradiation values are acceptable for many CCD applications. However, *Chang* [6.29] has also observed that the transfer efficiency in surface channel devices fabricated with this radiation-hard oxide was seriously degraded for doses greater than 10^5 rad (Si) (see Table 6.4). A severe increase in interface state trapping required the use of a 50 % bias charge to obtain a transfer efficiency of 0.992 after 10^6 rads (Si). The dark current density and

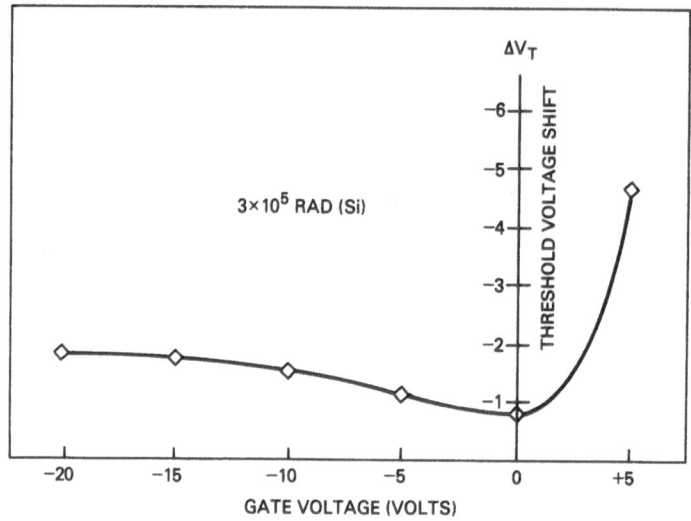

Fig. 6.17. Threshold voltage shift as a function of gate bias, illustrating the radiation sensitivity of the CCD hard oxide for positive gate bias during irradiation

threshold voltage shift after 10^6 rads (Si) were approximately equal to the values observed in the buried channel structure.

It should be noted that the pyrogenic CCD hard oxide process described above is insensitive to total dose radiation effects only for negative gate to channel bias (i.e., p-surface channel and n-buried channel). The flat-band shift for positive gate to channel bias, shown in Fig. 6.17, can be quite large.

Bulk silicon damage effects in neutron-irradiated radiation-hard devices do not differ from those in unhardened structures.

6.6 Irradiation Effects at Cryogenic Temperatures

CCDs are being considered for use in space and infrared imaging systems both as infrared detectors and as signal processors for IR focal plane arrays at low temperatures. However, several research groups, *Nielsen* and *Nichols* [6.30] and *Harari* et al. [6.31], reported that charge buildup in SiO_2 is more rapid for irradiation at 80 K than at 300 K. *Boesch* et al. [6.32] observed large flat-band voltage shifts after irradiation at 77 K in oxides which are radiation hard at room temperature. The flat-band voltage shifts at low temperature obtained by *Killiany* [6.33] for the CCD radiation-hard oxide are shown in Fig. 6.18. These large shifts render the devices unsuitable for most 77 K applications after a dose of 5×10^4 rads (Si). *Othmer* and *Srour* [6.34] recently found that the flat-band voltage shift in the CCD hard oxide during irradiation at 4.2 K is identical to the shift for 77 K irradiation.

Hughes [6.35] reported that an apparent immobilization of holes occurs in oxides at 74 K. Hence the holes generated in the oxide during a liquid nitrogen temperature irradiation are trapped almost immediately producing a nearly uniform density of positive charge in the oxide. The resultant flat-band voltage shift is proportional to the oxide thickness squared and has a value of -2.0 V per 10^4 rad (Si) for a 1000 Å oxide if all the generated holes are trapped uniformly. *Srour* and *Chiu* [6.36] observed that the fraction of the holes trapped is not strongly dependent on gate voltage polarity but is a function of the electric field strength in the oxide (see Fig. 6.19).

Harari et al. [6.31] used several techniques to anneal the excess flat-band voltage shift observed in devices irradiated at low temperatures. These include: photodepopulation of the traps, field-aided emission of holes from traps, and thermal annealing. *Boesch* et al. [6.37] observed that the annealing process in room-temperature radiation-hard oxides is accelerated for temperatures greater than 125 K (see Fig. 6.20).

Killiany [6.33] annealed the excess low-temperature irradiation threshold voltage shift observed in a room-temperature radiation-hard CCD by warming the device to room temperature. After recooling the CCD threshold voltage shift was approximately equal to the shift which would have been observed if the devices had been irradiated at 300 K and then cooled. The input gate threshold voltage shift for several irradiation-anneal cycles is shown in Fig. 6.21. The dotted lines indicate the threshold voltage shift during the thermal annealing process (approximately 20 min was required to warm the dewar). The first peak in Fig. 6.21 corresponds to a dose of 3×10^4 rads at 85 K. The second and third peaks are the shifts for incremental doses of 5×10^4 rads at 85 K. Normal CCD clock and bias voltages were applied to the device during the irradiation-annealing sequence. The residual shift at liquid nitrogen temperatures increased to -0.5 V after 1.3×10^5 rads as expected from 300 K irradiation results. A residual shift of up to -5 V might be expected for a typical nonhardened oxide subjected to the same irradiation-annealing cycle.

Srour et al. [6.38] have suggested several techniques [alternate insulators, MNOS (metal nitride oxide semiconductor) structures, and aluminum implanted oxides] for improving the low-temperature radiation behavior of MIS structures, and their use in CCD fabrication has been considered. However, *Saks* [6.39] has shown that the CCD compatible thin oxide MNOS approach offers considerable improvement in the radiation hardness of capacitors irradiated at 77 K. The flat-band voltage shifts observed in test capacitors having 1000 Å of silicon nitride (Si_3N_4) and 100 Å of silicon dioxide were -0.4 V and -2.2 V after a dose of 5×10^5 rads for an irradiation bias of ± 8 V, respectively. The small flat-band voltage shift is attributed to the escape of a large fraction of the radiation-generated holes from the oxide. *Boesch* and *McGarrity* [6.40] have determined that the holes generated in some oxides at 77 K have a mean free path on the order of 100 Å before being trapped. The relatively smaller shift for positive bias is attributed to trapping of the electrons generated in the oxide at the oxide-nitride interface or in the bulk of the nitride. The

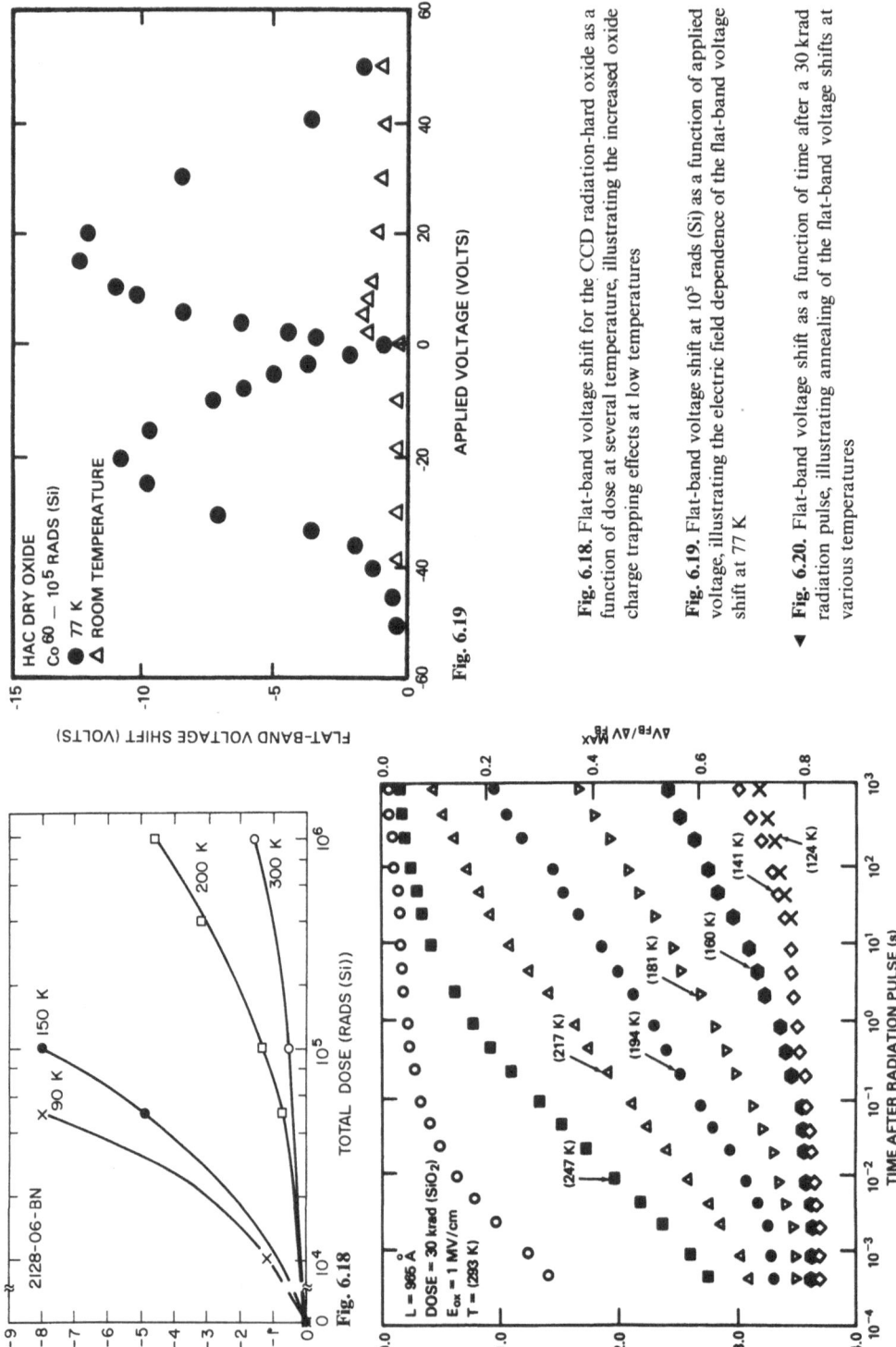

Fig. 6.18. Flat-band voltage shift for the CCD radiation-hard oxide as a function of dose at several temperature, illustrating the increased oxide charge trapping effects at low temperatures

Fig. 6.19. Flat-band voltage shift at 10^5 rads (Si) as a function of applied voltage, illustrating the electric field dependence of the flat-band voltage shift at 77 K

▼ **Fig. 6.20.** Flat-band voltage shift as a function of time after a 30 krad radiation pulse, illustrating annealing of the flat-band voltage shifts at various temperatures

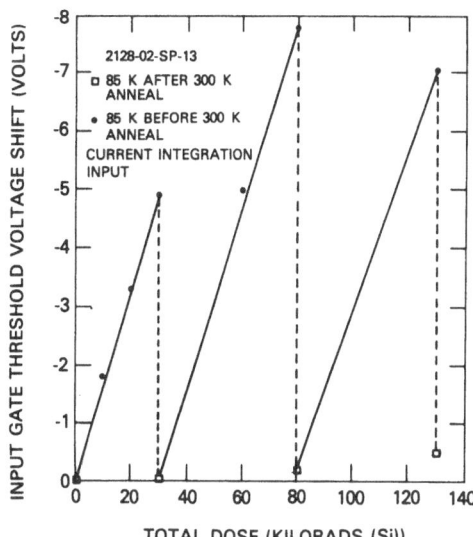

Fig. 6.21. Radiation-hard CCD input gate threshold voltage shift at 85 K as a function of dose and 300 K annealing (p-surface channel)

trapped electrons compensate the effect of the positively charged trapped holes in the oxide. An oxide thickness greater than 50 Å is required to exclude MNOS memory effects. The use of a thin oxide alone is not feasible due to yield and reliability (i.e., high fields) considerations. The large interface state densities usually observed in MNOS structures would require a buried channel CCD. Surface state dark current generation is not expected to be a problem for 77 K operation. The bias dependence of the threshold voltage shift indicates that a p-buried channel device would be the most radiation-hard structure.

The limited amount of experimental data available suggests that the permanent bulk silicon damage caused by fast neutrons in CCDs irradiated at 80 K and 294 K is similar. *Saks* et al. [6.17] reported that the energy and density of the $N-1$ trap level was independent of irradiation temperature (80 or 294 K) and irradiation bias. However, the transfer inefficiency in a buried channel device will change as the temperature is varied since the bulk trap emission time constant τ is temperature dependent. According to *Mohsen* and *Tompsett* [6.16] the electron emission time constant of a bulk level is given by

$$\tau^{-1} = \sigma_n v_{th} N_c \exp[-(E_c - E_t)/kT],$$

where σ_n is the electron capture cross section of the traps, v_{th} is the average electron thermal velocity, N_c is the density of states in the conduction band, and $(E_c - E_t)$ is the energy of the trap level below the conduction band. The shift of the periodic pulse transfer inefficiency curve to longer emission times with reduction in CCD operating temperature obtained by *Saks* [6.19] is shown in Fig. 6.22. At 80 K the emission time constant for the trap level $(N-3)$ which controls the transfer loss at 294 K has increased to a time much longer than 1 s. Consequently, these traps remain filled during the usual times involved in the

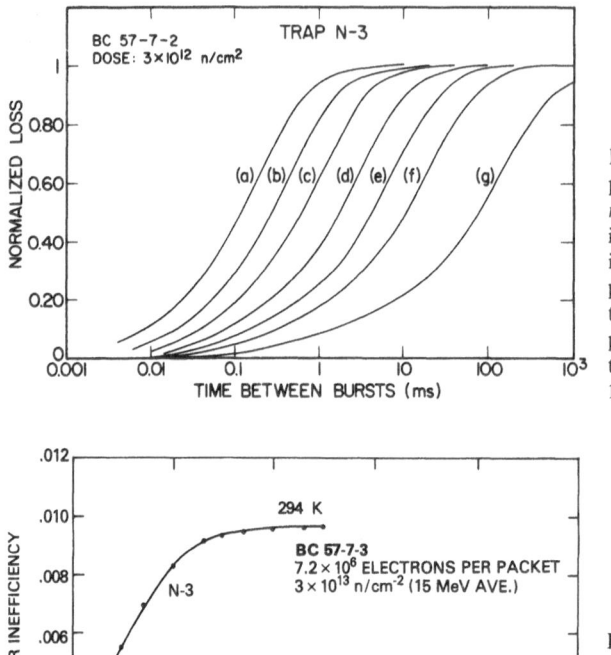

Fig. 6.22. Normalized periodic pulse data (i.e., $\varepsilon/\varepsilon_{max}$) for an n-buried channel device irradiated to 3×10^{12} n/cm², illustrating the shift in the periodic pulse curve to longer times [i.e.. position (a) to position (g)] as the temperature is changed from 262 to 199 K

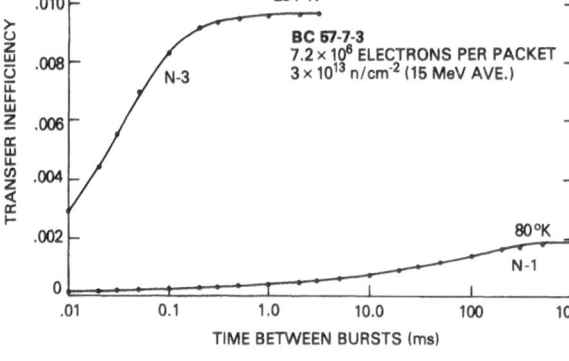

Fig. 6.23. Periodic pulse curves at 294 and 80 K for a CCD irradiated to 3×10^{13} n/cm². At 80 K the $N-1$ has a smaller density than the trap level effective at 294 K. Hence, the charge transfer inefficiency is smaller at 80 K

periodic pulse measurement and do not contribute to loss of signal charge in the CCD. The post-neutron irradiation transfer loss at 80 K is dominated by the lower density $N-1$ trap level. The reduced transfer inefficiency at 80 K is compared to the 294 K values in Fig. 6.23.

6.7 Summary

The previous sections of this chapter have presented on overview of the current state of the art of CCD radiation effects and hardening techniques. The key points presented are listed below.

1) CCDs are sensitive to both surface and bulk damage effects due to radiation. Devices fabricated using standard commercial gate oxide technologies are unable to satisfy the total dose requirements for most space and strategic requirements.

2) A buried channel device is the least radiation-sensitive structure for total ionizing dose effects.

3) The neutron-induced degradation of the charge-transfer efficiency is greater in buried channel devices than in surface channel devices.

4) Only a limited increase in the total dose radiation tolerance of CCDs can be achieved by means of structural and operational considerations alone. A radiation-hard oxide technology in addition to the optimized structure is required to satisfy system radiation requirements.

5) CDDs are extremely sensitive to transient upset effects. Increased tolerance can be achieved by thinning. Device burnout is prevented by current limiting the power supplies.

6) A megarad-hardened n-buried channel hard CCD technology has been developed and simple linear radiation-hard shift registers have been fabricated with the optimized structure and hard oxide technology.

7) Total ionizing dose effects are more severe for irradiation at 77 K. Ordinary room-temperature radiation-hardening techniques do not apply for irradiation at liquid nitrogen temperatures. However, the MNOS technique has been employed to solve this problem.

8) The degradation of CCD parameters in neutron-irradiated n-buried channel CCDs are less severe at 77 K than at 300 K.

References

6.1 B.L.Gregory, C.W.Gwyn: Proc. IEEE **62**, 1264 (1974)
6.2 F.Larin: *Radiation Effects in Semiconductor Devices* (Wiley, New York 1968)
6.3 R.J.Chaffin: *Microwave Semiconductor Devices: Fundamentals and Radiation Effects* (Wiley, New York 1973)
6.4 C.H.Sequin, A.M.Mohsen: IEEE J. SC-**10**, 81 (1975)
6.5 J.M.Killiany, W.D.Baker: "Limitations of a Threshold-Insensitive CCD Input Technique in a Total Dose Radiation Environment," in 1975 Intern. Conf. Applic. CCDs Proc., pp. 369–374
6.6 D.F.Barbe, J.M.Killiany, H.L.Hughes: Appl. Phys. Lett. **23**, 400 (1973)
6.7 J.M.Killiany, W.D.Baker, N.S.Saks, D.F.Barbe: IEEE Trans. NS-**21**, No. 6, 193 (1974)
6.8 J.E.Carnes, W.F.Kosonocky: Solid State Technol. **17**, 67 (1974)
6.9 G.A.Hartsell: "Radiation Hardness of Surface and Buried Channel CCDs," in 1975 Intern. Conf. Applic. CCDs Proc., pp. 375-382
6.10 G.A.Hartsell, D.A.Robinson, D.R.Collins: "Effects of Ionizing Radiation on CCDs," in 1975 Symp. CCD Tech. Sci. Imaging Applic. Proc., pp. 220–227
6.11 R.A.Williams, R.D.Nelson: IEEE Trans. NS-**22**, 2639 (1975)
6.12 T.C.May, M.H.Woods: "A New Physikal Mechanism for Soft Errors in Dynamic Memories", in 1978 Reliability Phys. Symp. Proc., pp. 33-40
6.13 J.C.Pickel, J.T.Blandford, Jr: IEEE Trans. NS-**25**, 1166 (1978)
6.14 W.Shedd, B.Buchanan: IEEE Trans. NS-**23**, 1636 (1976)
6.15 G.W.Autio, M.A.Bafico: Infrared Phys. **15**, 249 (1975)
6.16 A.M.Mohsen, M.F.Tompsett: IEEE Trans. ED-**21**, 701 (1974)
6.17 N.S.Saks, J.M.Killiany, W.D.Baker: "Effects of Neutron Irradiation on the Characteristics of a Buried Channel CCD at 80 and 295 K," in 1976 NASA-JPL Conf. CCD Tech. Applic. Proc., pp. 110–114

6.18 E.C.Smith, D.Binder, P.A.Compton, R.I.Wilbur: IEEE Trans. NS-**13**, No. 6, 11 (1966)

6.19 N.S.Saks: IEEE Trans. NS-**24**, 2153 (1977)

6.20 J.W.Walker, C.T.Sah: Phys. Rev. B**7**, 4587 (1973)

6.21 O.L.Curtis, J.R.Srour: IEEE Trans. NS-**20**, No. 6, 193 (1973)

6.22 H.J.Stein, R.Gereth: J. Appl. Phys. **31**, 2890 (1968)

6.23 J.R.Srour, S.C.Chen, S.Othmer, R.A.Hartmann: IEEE Trans. NS-**25**, 1251 (1978)

6.24 O.L.Curtis: IEEE Trans. NS-**13**, No. 6, 33 (1966)

6.25 J.E.Carnes, A.D.Cope, L.R.Rockett: "Effects of Radiation on Charge-coupled Devices," Final Report, Contract No. RADC-TR-76-285, RCA Laboratories (1976)

6.26 C.P.Chang, K.G.Aubuchon: "CCD Radiation Hardening," Final Report, Contract Number N00173-77-C-0158, Hughes Aircraft Company, Newport Beach, CA June (1978)

6.27 S.Mayo, K.F.Galloway, T.F.Leedy: IEEE Trans. NS-**23**, 1875 (1976)

6.28 C.P.Chang: IEEE Trans. NS-**25**, 1454 (1978)

6.29 C.P.Chang: IEEE Trans. NS-**23**, 1639 (1976)

6.30 R.L.Nielsen, D.K.Nichols: IEEE Trans. NS-**20**, No. 6, 319 (1973)

6.31 E.Harari, S.Wang, B.S.H.Royce: J. Appl. Phys. **46**, 1310 (1975)

6.32 H.E.Boesch, Jr., F.B.McLean, J.M.McGarrity, G.A.Ausman, Jr.: IEEE Trans. NS-**22**, 2163 (1975)

6.33 J.M.Killiany: IEEE Trans. NS-**24**, 2194 (1977)

6.34 J.R.Srour, S.Othmer, S.C.Chen, R.A.Hartman: "Investigation of the Basic Mechanisms of Radiation Effects on Semiconductor Devices Used in Electro-Optical Sensor Applications", Final Report, Contract Number DNA 001-78-C-0028, Northrop Corporation, Palos Verdes Peninsula, CA, August (1979)

6.35 R.C.Hughes: Appl. Phys. Lett. **26**, 436 (1975)

6.36 J.R.Srour, K.Y.Chiu: IEEE Trans. NS-**24**, 2141 (1977)

6.37 H.E.Boesch, Jr., J.M.McGarrity, F.B.McLean: IEEE Trans. NS-**25**, 1012 (1978)

6.38 J.R.Srour, S.Othmer, O.L.Curtis, Jr., K.Y.Chiu: IEEE Trans. NS-**23**, 1513 (1976)

6.39 N.S.Saks: IEEE Trans. NS-**25**, 1226 (1978)

6.40 H.E.Boesch, Jr., J.M.McGarrity: IEEE Trans. NS-**23**, 1520 (1976)

Subject Index

Acoustic Surface Waves

Editor: A. A. Oliner

1978. 198 figures, 16 tables. XI, 331 pages
(Topics in Applied Physics, Volume 24)
ISBN 3-540-08575-0

Contents:
Types and Properties of Surface Waves. –
Principles of Surface Wave Filter Design. –
Fundamentals of Signal Processing Devices.
Waveguides for Surface Waves. Materials and
Their Influence on Performance. – Fabrication Techniques for Surface Wave Devices.

Integrated Optics

Editor: T. Tamir

2nd corrected and updated edition.
1979. 99 figures, 11 tables.
XV, 333 pages
(Topics in Applied Physics, Volume 7)
ISBN 3-540-09673-6

Contents:
T. Tamir: Introduction. – *H. Kogelnik:* Theory
of Dielectric Waveguides. – *T. Tamir:* Beam
and Waveguide Couplers. – *J. M. Hammer:*
Modulation and Switching of Light in
Dielectric Waveguides. – *E. Zernike:* Fabrication and Measurement of Passive Components. – *E. Garmire:* Semiconductor Components for Monolithic Applications. – *T. Tamir:*
Recent Advances in Integrated Optics. –
Additional References with Titles. – Subject
Index.

Optical and Infrared Detectors

Editor: R. J. Keyes

1977. 115 figures, 13 tables. XI, 305 pages
(Topics in Applied Physics, Volume 19)
ISBN 3-540-08209-3

Contents:
R. J. Keyes: Introduction. – *P. W. Kruse:* The
Photon Detection Process. – *E. H. Putley:*
Thermal Detectors. – *G. D. Long:* Photovoltaic and Photoconductive Infrared Detectors. – *H. R. Zwicker:* Photoemissive Detectors. – *A. F. Milton:* Charge Transfer Devices
for Infrared Imaging. – *M. C. Teich:* Nonlinear
Heterodyne Detection.

X-Ray Optics

Applications to Solids

Editor: H.-J. Queisser

1977. 133 figures, 17 tables. XI, 227 pages
(Topics in Applied Physics, Volume 22)
ISBN 3-540-08462-2

Contents:
H.-J. Queisser: Introduction: Structure and
Structuring of Solids. – *M. Yoshimatsu,
S. Kozaki:* High Brilliance X-Ray Sources. –
E. Spiller, R. Feder: X-Ray Lithography. –
U. Bonse, W. Graeff: X-Ray and Neutron
Interferometry. – *A. Authier:* Section Topography. – *W. Hartmann:* Live Topography.

Springer-Verlag Berlin Heidelberg New York

A. H. Eschenfelder

Magnetic Bubble Technology

1980. 271 figures, 8 tables. Approx. 360 pages
(Springer Series in Solid-State Sciences,
Volume 14)
ISBN 3-540-09822-4

Contents:
Introduction to Magnetic Bubbles. – Static
Properties of Magnetic Bubbles. – Dynamic
Properties of Magnetic Bubbles. – Basic
Permalloy-Bar Bubble Devices. – Other
Bubble Device Forms. – Bubble Materials.
– Device Chip Fabrication. – Chip Packaging. – Applications. – Future Prospects. –
References. – Subject Index.

T. Kohonen

Content-Adressable Memories

1980. 123 figures, 35 tables.
Approx. 400 pages
(Springer Series in Information Sciences,
Volume 1)
ISBN 3-540-09823-2

Contents:
Associative Memory, Content Addressing,
and Associative Recall. – Content Addressing
by Software. – Logic Principles of Content-
Addressable Memories. – CAM Hardware. –
The CAM as a System Part. – Content-
Addressable Processors. – References. –
Subject Index.

R. H. Kingston

Detection of Optical and Infrared Radiation

1978. 39 figures, 2 tables. VIII, 140 pages
(Springer Series in Optical Sciences,
Volume 10)
ISBN 3-540-08617-X

Contents:
Thermal Radiation and Electromagnetic
Modes. – The Ideal Photon Detector. –
Coherent or Heterodyne Detection. –
Amplifier Noise and Its Effect on Detector
Performance. – Vacuum Photodetectors. –
Noise and Efficiency of Semiconductor
Devices. – Thermal Detection. – Laser
Preamplification. – The Effects of Atmospheric Turbulence. – Detection Statistics.
Selected Applications.

Noise in Physical Systems

Proceedings of the Fifth International
Conference on Noise, Bad Nauheim,
Fed. Rep. of Germany, March 13–16, 1978
Editor: D. Wolf

1978. 182 figures, 5 tables. X, 337 pages
(Springer Series in Electrophysics, Volume 2)
ISBN 3-540-09040-1

Contents:
Noise in Semiconductor Devices. – Hot
Carrier Noise. – 1/f-Noise. – Noise in
Magnetic Materials. – Noise in Super-
conductors and Superconducting Devices. –
Noise Measuring Techniques. – Theory.

Springer-Verlag Berlin Heidelberg New York